I0059604

OUR GENETIC DESTINY

OUR GENETIC DESTINY

Understanding the Secret of Life

Amil Shah M.D.

Hounslow

Our Genetic Destiny: Understanding the Secret of Life

Copyright © 1996 by Amil Shah M.D.

All Rights Reserved. No part of this publication may be reproduced, stored in a retrieval system, or transmitted in any form or by any means, electronic, mechanical, photocopying, recording, or otherwise (except brief passages for the purposes of review) without the prior permission of Hounslow Press. Permission to photocopy should be requested from the Canadian Reprography Collective.

Hounslow Press
A member of the Dundurn Group

Publisher: Anthony Hawke
Editor: Liedewy Hawke
Designer: Sebastian Vasile and Evelyn David
Interior Illustrations: Amil Shah
Printer: Webcom
Front cover photograph: Automated DNA sequencing using fluorescently labelled nucleotides. [Courtesy: Dr. Doug Horsman and Sheryle Adomat]

Canadian Cataloguing in Publication Data

Shah, Amil, 1949-
  1. Our genetic destiny: understanding the secret of life

Includes bibliographical references
ISBN 0-88882-189-1

1. Genes.   I. Title

QH447.S43 1996   574.87'322   C96-930604-0

Publication was assisted by the **Canada Council**, the **Book Publishing Industry Development Program** of the **Department of Canadian Heritage**, and the **Ontario Arts Council**.

Care has been taken to trace the ownership of copyright material used in this book. The author and the publisher welcome any information enabling them to rectify any references or credit in subsequent editions.

Hounslow Press
2181 Queen Street East
Suite 301
Toronto, Ontario, Canada
M4E 1E5

Hounslow Press
73 Lime Walk
Headington, Oxford
England
OX3 7AD

Hounslow Press
250 Sonwil Drive
Buffalo, NY
U.S.A. 14225

CONTENTS

ACKNOWLEDGEMENTS

Writing this book was made immensely more enjoyable through the encouragement of my wife, Colleen, and our children, Justin, Alia, and Chantal. I was also fortunate to have the enthusiastic support of a few friends, especially John McDonnell, Lillian Starr, Kathleen Martens, and Al Hurwitz.

I wish to thank Ferguson Neville for his improvements to my drawings, and Dr. Doug Horsman and Sheryle Adomat for kindly providing the photograph for the front cover.

For his confidence in my work and for the efforts of his staff at Hounslow Press, I tender my thanks to Tony Hawke.

PROLOGUE

At the heart of life lie our genes. They tell our cells what to do, when to develop, and what to become. We go through life oblivious to most of this. Our cells are born with a complete set of instructions: our heart muscle cells contract and relax with unfailing regularity to pump blood throughout the body; our pancreatic cells produce the enzymes to digest the food we eat; our brain cells respond with precision and rapidity to external stimuli. All this, and more, goes on, governed by our genes, without ever a personal word from us. It is a fantastic arrangement, built-in and generally infallible.

We get along in life this way because we are endowed with genetic instructions inherited from our parents and passed on to our children. All the genetic information is stored like a written document on long, chain-like molecules of deoxyribonucleic acid, thankfully abbreviated to DNA. Strung through every cell, DNA is the greatest single achievement of Nature. It makes us who we are.

DNA has been around for a long, long time, nearly four thousand million years or so. Over this time, it has endured, but not without some changes. The original molecule has been modified, embellished, mutated, improved. The astonishing diversity of living things around us is its legacy, the real marvel of DNA. The overarching goal of modern biology is to figure out how the molecule of DNA has the capacity to organise living things, from the bland lives of primitive bacteria to the firing neurones of the human mind.

The era of modern biology started only recently, in 1953, when the helical molecule of DNA was revealed in all its mystery. This discovery was greeted with an outburst of enthusiasm and bewilderment. It was, no doubt about it, an epoch-making event. It was not so much the elucidation of the structure of the molecule that caused this excitement as the realisation that DNA is the vehicle of heredity; in it are all the instructions for life and for living. The magnitude of this discovery was such that some were tempted to declare that the biological sciences had accomplished all there was to know. With the description of the central molecule of life, there could not be much more to do. From here on, it was largely a problem of finding out how DNA tells the cell what to do. Once the loose ends were tied up, it would all be over.

But exactly how DNA orchestrates the activities of the cell is what the new biology is all about. In the past few years, DNA has become almost imponderably complex, more subtle and intricate than we ever imagined. As we forge a path towards the cell's nucleus, we find it filled with strange parts whose functions are still beyond explanation. New questions constantly crop up, and the answers keep piling up. As more is understood, however, general schemes, beautifully simple but absorbingly fascinating, begin to emerge.

The considerable progress in the recent past has changed how we look at biology. No longer is it regarded as a staid subject, just a long and tedious list of classification of different species, which high school students were obliged to memorise, but a dynamic discipline, replete with the unexpected. Each new turn brings into view a whole new vista to gaze upon in splendid awe. Biology has become, in the most literal sense of the phrase, a subject full of life.

Living things do not pass on a simulacrum of themselves to the next generation. Instead, they provide it with the necessary genetic instructions to build a progeny in their image. The nature of this information and how it is kept are now known. Genes are the guardians of life's heritage.

Two central questions immediately come to mind. First, how is the information accurately passed down from one generation to the next; and, second, how is it used by the organism for living? The answers to these questions were worked out in broad outlines through the diligent efforts of biologists in the middle of this century, and led to the enunciation of a Central Dogma: DNA is the hereditary material, and it provides the blueprint for the construction of proteins, the molecular purveyors of cellular function. This dictum framed our thinking on the subject for a long time.

Living organisms turn out to be incredibly complex. Intriguing twists to the more traditional, simplistic models of DNA's behaviour, embodied in the Central Dogma, keep on coming. These discoveries have gone on quietly, unnoticed by the general public. Perhaps, this is because the foundation upon which biological principles rest has not been disturbed.

Not yet, at least.

Contrary to expectations, genes are anything but static. They amplify and duplicate themselves; they expand and contract like accordions; they leave the chromosome and live as separate entities for a while; they leap from one chromosome to another; they reshuffle to create new images of themselves. Proteins can sometimes be specified by genes that, in the conventional sense, do not exist.

The new heresies being discussed are leading to major revisions in genetics. Already, biologists are waking up to the fact that DNA is not quite the predictably stable molecule, following a preset schedule, that was taken for granted just a short while ago. Afterall, for a molecule responsible for the genetic continuity of a species from one generation to the next, change was the last thing expected.

For a molecule born in a bolt of lightning when the Earth first swung into orbit some 4.7 billion years ago, it has come a long way. The first cell came on the scene not long afterwards, and spent most of its time learning the business of living. And then, in a spectacular moment, 600 million years ago, it burst forth in an awesome explosion of new forms of life. In every cell on the Earth today are bits of that original DNA.

Even today, genes continue to adapt to new challenges. What we are learning is reshaping how we look at life, and, indeed, how we look at ourselves.

It is a wonderful parade.

How manifold are Thy works, O Lord!
In wisdom hast Thou made them all;
The Earth is full of Thy creatures.
Psalm 104 verse 24

From so simple a beginning endless forms most
beautiful and most wonderful have been, and are
being evolved.
The Origin of Species - Charles Darwin

The world is so full of a number of things,
I'm sure we should all be as happy as kings.
A Child's Garden of Verses - Robert Louis Stevenson

CHAPTER 1

THE DANCE OF LIFE

Brunn, February 1865.

Almost all of the forty members of the Brunn Society for the Study of Natural Sciences gathered to hear one of its founding members deliver a paper. It was widely known amongst them that Father Gregor Johann Mendel was working on hybridisation of plants. This was Mendel's first public presentation of his findings.

The road to Brunn — now Brno — was not an easy one for Mendel. He was born in 1822, the only son of Silesian peasants who lived in the Austrian hamlet of Heitzendorf. Mendel was a good student who so impressed the local schoolmaster that he pleaded with the boy's father to send him to Troppau to further his studies. Despite the financial hardship of this decision, he did so. Later, Mendel went to the Augustinian monastery at Brunn, and then to the University of Vienna. He studied the sciences, and his greatest love was for botany. Mendel followed the contemporary debates on plant evolution, and became interested in how certain traits were passed from one generation to the next. This proved too *avant garde* for his examiners at Vienna, who disapproved of his ideas. Mendel never passed his examination to be a teacher. He disputed some point or the other with his examiners and would not give in. He flunked.

Mendel returned to Brunn where he pursued his interest in the inheritance of different traits in plants. He experimented with peas for seven years in a small strip of garden, 120 feet by 120 feet, at the back of a monastery. Here, the first and most important step in solving the riddle of heredity would take place. Mendel knew that some pea plants bred true for certain traits. For example, some plants produced only yellow peas, others only green peas. When he cross-bred these two strains, he expected to get an equal amount of yellow and green peas. Instead, to his surprise, all the peas were yellow; the green peas seemed to have vanished. He then planted the yellow seeds from this hybrid expecting to get only yellow peas. When the plants grew up, there was a mixture of yellow and green peas. The green peas had reappeared in the second generation. He tried this with other traits. He cross-bred round and wrinkled peas, and the first cross produced only round peas. Plant these, and among the round peas of the second generation were some wrinkled peas.

These observations defied logic. Mendel surmised that some traits like yellow colour were stronger, and he called them dominant; others like green colour were weaker or recessive. Yet, the weak traits always reappeared in the second generation of the hybrid peas. Mendel kept careful records of his experiments, and after two years of crossing pea plants, he made a brilliant observation. The ratio of the strong to weak traits was always three to one.

In one of the hybrid harvest, he got 8,023 peas, of which 6,022 were yellow and 2,001 were green: a 3:1 ratio.

Much to his credit, Mendel was able to deduce the meaning of this ratio, thanks to his understanding of mathematics. He assumed that each plant inherited two factors, one from each parent, that were transmitted in the germ cells. He labelled the dominant factor with the capital letter A, and the weaker factor with the small a. The yellow plants that bred true contained two factors, and were AA; the green plants were aa. The germ cells — the male pollen and female egg cell — of the yellow plant each contained one A factor; likewise, the germ cells of the green plant contained one a factor.

Mendel worked out the composition of the plants following cross-breeding of yellow and green plants. This is shown in Figure 1a. The first cross led to hybrid plants Aa. As A was dominant, all the peas were yellow. The germ cells of this first generation were either A or a. By allowing these to cross-fertilise themselves, he noted the ratio 3:1, as is depicted in Figure 1b. The second generation of plants consisted of a pure yellow plant (AA), two hybrid yellow plants (Aa), and a pure green plant (aa).

Mendel tested his hypothesis with different traits. The first generation always yielded plants with the dominant trait, while the second generation gave rise to plants in the 3:1 ratio. This was a significant finding, and even more impressive was Mendel's acumen to figure out the reasons for it.

This would have been enough for most, but not Mendel. He took it one step further. He asked how would two different traits assort. He mixed peas with different colours — yellow or green — and smoothness — round or *a. First generation of cross between yellow and green peas*

PARENTS: AA   or   aa
yellow   *green*

GERM CELLS: A or A   a or a

	A	A
a	Aa	Aa
a	Aa	Aa

Result: All peas are yellow

b. Second generation of cross between hybrid peas

PARENTS   Aa   Aa

GERM CELLS   A or a   A or a

	A	a
A	AA	Aa
a	Aa	aa

Result: Three peas are yellow and one is green

Figure 1. Results of Mendel's experiments of crossing peas of different colours.

wrinkled. He predicted that they would grow in the ratio 9:3:3:1 — nine that were yellow and round, three that were yellow and wrinkled, three that were green and round, and one that was green and wrinkled. Figure 2 explains how he arrived at these numbers. Mendel's predictions were correct. Add a third trait, and the ratio became 27:9:9:3:9:3:3:1. Again, the results of the breeding experiments confirmed his calculations.

	AB	Ab	aB	ab
AB	AB AB	Ab AB	aB AB	ab AB
Ab	AB Ab	Ab Ab	aB Ab	ab Ab
aB	AB aB	Ab aB	aB aB	ab aB
ab	AB ab	Ab ab	aB ab	ab ab

A = yellow   a = green
B = round   b = wrinkled

Results:			
	yellow round	=	9
	yellow wrinkled	=	3
	green round	=	3
	green wrinkled	=	1

Figure 2. Results of experiments of crossing pea plants with different colours and textures.

Mendel had finally unlocked the secrets of inheritance. First, he suggested that hereditary traits were controlled by factors, now called genes, that were passed on to each successive generation. Second, to explain the disappearance and reappearance of a particular trait in a predictable manner, he correctly deduced that each trait was produced by two genes, with one being contributed by each parent and recombined at fertilisation of the male and female germ cells. Third, when two contrasting genes were present, only one was expressed.

On a cold winter evening in 1865, Gregor Mendel presented his findings to the Brunn Society. No one said a word. The mixing of mathematics and botany was just not done. Mendel's grand experiment was all for naught. The Society extended to him the courtesy of publishing his work in their proceedings the following year. Copies were duly sent to universities and scientific societies around Europe, where they collected dust for decades. Mendel did try to seek the help and advice of Karl von Nageli, a contemporary botanist of some renown, but again he was ignored.

Mendel was ahead of his time. By the turn of this century, however, chromosomes were recognised as a basic feature in cells, and the new science of genetics was born. The world was finally ready for Mendel. There are few instances in science where we can trace the origin of an idea so clearly as Mendel's principles of heredity. He gave us the concept of genes, and in so doing charted the course that biologists were to follow to understand life.

And what is life?

No one observing the familiar sight of young children at play among the driftwood on a beach would have any trouble defining what is living and what is not. It seems so obvious that we seldom give the subject any thought. Yet, life is a most remarkable phenomenon when we do pause to think about it.

Adding to the wonder is that once we delve into the matter, we come face to face with an irreducible paradox: Living things are made of inanimate molecules. Six elements — carbon, hydrogen, oxygen, nitrogen, phosphorus, and sulphur — are the chief ingredients of life. A few other elements, such as calcium, chlorine, potassium, sodium, and magnesium, are also vital, while a collection of sixteen more in trace amounts play a role in some species. The interesting thing is that only a small number of the 90 naturally occurring elements on Earth go into the making of living creatures. There is nothing unique about these; they are ubiquitous and they are abundant.

Of the different elements, carbon occupies a central position due to its special atomic structure. To understand this, we must examine certain properties of elements. All fall into one of three categories. There are gases, such as hydrogen, helium, oxygen; metals, such as sodium, iron, potassium; and non-metallic solids, such as carbon, sulphur, phosphorus.

In 1869, Dmitri Mendeleev developed a "periodic table," in which he grouped the elements that showed a strong family resemblance together. In the modern version of his table, the elements are arranged into six common groups, based on their atomic weights. Curiously, the sizes of the groups are 2,8,8,18,18, and 32. At first, physicists could make nothing of these numbers, but therein was a fundamental property of the elements.

This was worked out by the physicist Neils Bohr. According to his theory, atoms are made of three kinds of particles: protons, which carry a positive charge; neutrons, which carry no charge; and electrons, which carry a negative charge exactly opposite to that of protons. Protons and neutrons are found in the centre of the atom, forming the nucleus, whereas the electrons orbit around the nucleus at specific distances or shells.

By and large, the atomic weight of any atom is determined by the number of protons and neutrons, each of which has an almost equal weight; the electrons contribute a negligible amount to the atom's weight. The simplest atom is hydrogen, made up of one proton and one electron. It is assigned an atomic weight of 1. Next in the periodic table is helium with two protons, two neutrons, and two electrons; its atomic weight is 4. Further along is carbon with six protons, six neutrons, and six electrons; atomic weight 12.

Back to Mendeleev's strange set of numbers. This turns out to be the number of electrons in the various shells around the nucleus. The innermost shell accommodates two electrons, the next shell eight, the third out eight, and so on in the series. The electrons are important to the behaviour of the elements. How one element interacts with another depends on the number of electrons in its outermost shell, as these react most easily with the electrons in the other element. For example, sodium has 11 electrons, arranged in successive shells of two, eight, and one. The third shell out has a single electron in orbit. Chlorine has 17 electrons in shells of two, eight, and seven. Its outermost shell has seven electrons; to be "filled" it needs another electron, so one is missing. The single electron in the outermost shell of sodium can be donated or shared with the seven in chlorine's outermost shell. The result of this sharing is to fill the respective shells of the two elements. Full shells are chemically stable. Hence, sodium and chlorine atoms readily interact to form a stable compound, sodium chloride or common table salt.

What about carbon? Carbon has six electrons, two in the first shell and four in the outer shell. It needs eight to form a full outer shell, so it is missing four electrons. Another way to look at it is that it can donate its four outer electrons, and this makes it unique. This peculiarity of carbon makes its chemistry more complicated than any other element. It permits a wide range of stable combinations, but most importantly, it allows carbon atoms to react with other carbon atoms to form complex structures. Without these, life as we know it would not be possible.

The simplest compounds formed by carbon are the hydrocarbons, composed of carbon atoms and hydrogen atoms (Figure 3). A single carbon atom can bind to four hydrogen atoms to form a gas, methane. Next in complexity is ethane, made of two carbon atoms linked to each other and surrounded by six hydrogen atoms. This is followed by propane with a row of three linked carbon atoms with eight hydrogen atoms. Increasingly longer chains can be made, but more important for life than chains of carbon atoms are closed rings. The simplest is the benzene ring, composed of six carbons and six hydrogens. The benzene ring is a vital unit, since it allows the building of the more complex, stable compounds found in living organisms.

The discovery of the benzene ring was at first difficult to explain. How could six carbon atoms combine with just six hydrogen atoms to form a stable compound? A simple chain of six carbon atoms would require 14

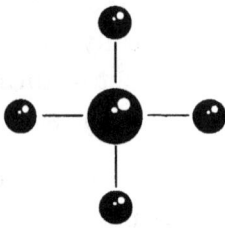

Methane

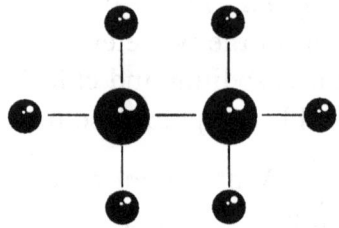

Ethane

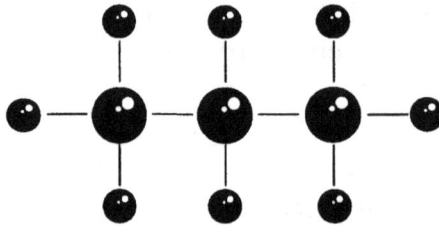

Propane

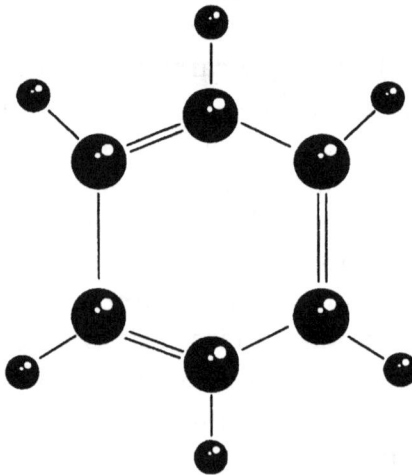

Benzene

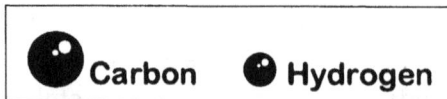

Figure 3. Chemical structure of simple hydrocarbons.

hydrogen atoms. The closed ring structure was proposed by Fredrick Kekule, who, so the story goes, conceived of this model in a dream of a snake eating its own tail.

With carbon as the central element, about thirty essential compounds are built: twenty amino acids; five nucleobases; two sugars; glycerol; an amino alcohol, choline; and a fatty acid, palmitic acid. These are the molecules of life.

Stripping down life — and indeed living — to such a stark chemical definition can be jolting. We seem more comfortable to inject mysticism into it — or at least get mysticism out of it. The reason is evident. Down through the ages, the whole subject has been shrouded in mystery. Each culture sought to explain life in terms of what was currently known to it. So when, in this modern age of biology, Francis Crick and James Watson gave us the double-helix, it was inevitable that we came to see life as a constant flow of information coded in molecules. This is in no way meant to detract from the marvel of it all. There is nothing to touch the spectacle. Far from a robot carrying out specific, predictable tasks, the living cell is an unimaginably complex, yet perfectly organised, invention of Nature.

Here, then, is the paradox. We are, all living creatures, made of simple elements abundantly scattered around the Earth. Nonetheless, we are capable of complicated, co-ordinated feats, even in the simplest, single-celled member of the family. Solving this paradox has long been the quest of biologists, who share the quiet passion to find a grand unifying principle of life, much like those for chemistry and physics. We lack this in biology. We continue to make observations and connect one fact with another, making progress here and there, but unable so far to put all the pieces together into a wonderful whole.

The effort in this quest received a major boost in the 1950s following the discovery of the master molecule of life, DNA. Until then, we ascribed the process of life to one mysterious agent or another, called by various names. By the end of the eighteenth century, the agent was a "vital force." This was perceived as a central force that permeated the body, in the muscle and in the nerve, in each and every tissue. The force endowed each organ with some specific attribute: sentiment, intuitive perception, and intellect.

The concept of vitalism became prominent because it filled a vacuum in the complete understanding of living things. Today, life is seen as having three components: matter, energy, and information. Earlier biologists had recognised the flow of matter, but lacking any knowledge of energy and information, they put forward the idea of a vital force. Now, with a detailed understanding of the chemistry of the cell, vital force is replaced by energy. Vitalism is redefined by the biochemical reactions that go on all the time when cellular substances react with one another to produce energy.

With the discovery of DNA, the information system of living things was established. DNA forms the underpinning of life. It is, beyond doubt, a triumph of Nature, lavish in its capacity to respond to any contingency, yet at its finest in managing the routine details of daily living.

The business of life is carried out in the cell, the elemental unit of living organisms, first described in the seventeenth century. Using a crude microscope made of a single magnifying lens, Robert Hooke looked at a slice of cork in 1665 and described a little honeycomb of holes. It reminded him of the cells of a monastery, and he called the tiny holes just that — cells. What he was looking at were the dried-up remnants of plants, but his was the first description of this most basic unit.

As better and more powerful microscopes became available, biologists began to focus on the fine details of the cell, which led eventually in the early part of this century to the recognition of the crucial role of its chromosomes in heredity. Until then, chromosomes were an oddity, but through the pioneering efforts of Theodor Boveri, attention was turned to understanding heredity: how certain traits, such as eye colour, might be passed on in a regular manner from parents to their offspring.

The chromosomes are cramped into a tiny space within the cell known as the nucleus (Figure 4). The nucleus is readily identified as a compact mass called chromatin, because it can be stained by special dyes. When a cell divides, however, the individual chromosomes come into the limelight as discrete thread-like particles that distribute themselves in precisely equal parts to the two daughter cells; each new cell that is created inherits an equal share of the family genetic heirloom.

The discovery of chromosomes rekindled interest in Mendel's factors. They provided the physical basis for inheritance, and the striking parallel between Mendel's factors and chromosomes could not be overlooked. Chromosomes are present in pairs, one set from each parent, which pass unchanged to the offspring. Each chromosome of a pair can be passed on independent of the members of other pairs, allowing for the various mixes of parental genetic traits in the offspring. Mendel was, at long last, vindicated.

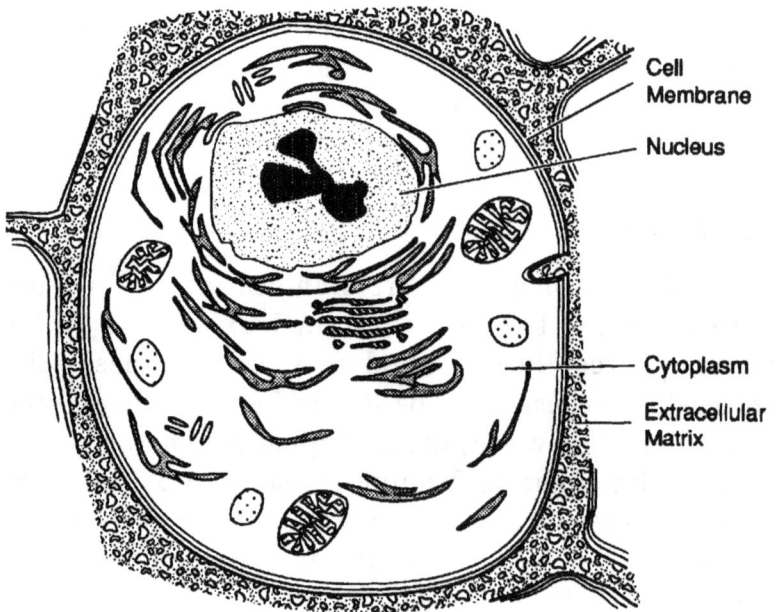

Figure 4. Structure of a living cell.

Along the chromosome are genes, arranged one after the other, in a fixed, linear order. They convey information from one generation to the next, information the cell needs to function. It did not take long for biologists to realise that genes carry the instructions to make proteins, molecules that come in a large variety of sizes and shapes, designed to sustain the pulse of life.

The importance of this connection between genes and proteins was enough to engage the minds of biologists for some time. At first, they assumed that genes must be as complex and diverse as proteins. In fact, however, the role of the gene is simply to provide the information to craft the protein molecules. This information is in the form of a code, enciphered in DNA.

A large molecule, DNA is formed from great numbers of smaller molecules. There are four bases, designated A, G, T, and C, for adenine, guanine, thymine, and cytosine; a sugar, deoxyribose; and phosphate. Despite the simple chemical composition, DNA molecules control the activities of every living thing; in them, all the instructions for living are kept on file.

DNA is the genetic material of all living organisms. Everything alive, people, plants, earthworms, birds, is run by DNA. It is the common thread that weaves itself throughout the fabric of life. In some viruses, the genetic material is ribonucleic acid or RNA, an alternative molecule that is only slightly different, although in these circumstances, it fulfils the same function as DNA. In RNA, the same bases are used, except that thymine is replaced by the base uracil (U), which is very similar chemically.

The fantastic geometry of DNA was worked by Watson and Crick at the University of Cambridge. The molecule is made of two strands lying side by side like a zipper, and held together by hooks — bases — which are attached to long ribbons of alternating molecules of sugar and phosphate. The two strands twist around each other to form a double helix (Figure 5). The bases are on the inside, and they pair up in a strict manner: A to T, and G to C. Adenine and guanine are large, while thymine and cytosine are smaller, so that each pair consists of a large base and a small base, providing a dyadic symmetry. The dimensions of the double helix are such that there is insufficient space to permit the two large bases to pair up, and would allow too much space between the two smaller bases.

The strands of the double helix are not identical, but are complementary to each other. This is the result of the strict pairing of the bases. Each of the two pairs can be inserted in the double helix in one of two ways: A = T or T = A, and C = G or G = C. The base pairs can exist in any sequence along the DNA, so that by varying the order in which the bases occur, different DNA molecules are created.

The alignments of the bases within the DNA molecules may be different in different species, but the molecules themselves are fundamentally the same substance. In the deep recess of our cells, out of sight, the DNA which modulates our brain waves is basically the same as that which co-ordinates the flapping of a hummingbird's wings or the beating of a bacterium's flagellae. We have more close cousins, all over the place, than we ever imagined.

Within the confines of the nucleus, DNA is wrapped around special proteins to form a compact mass. The need for such tight packing is evident from the length of the molecule, which when extended would vastly exceed

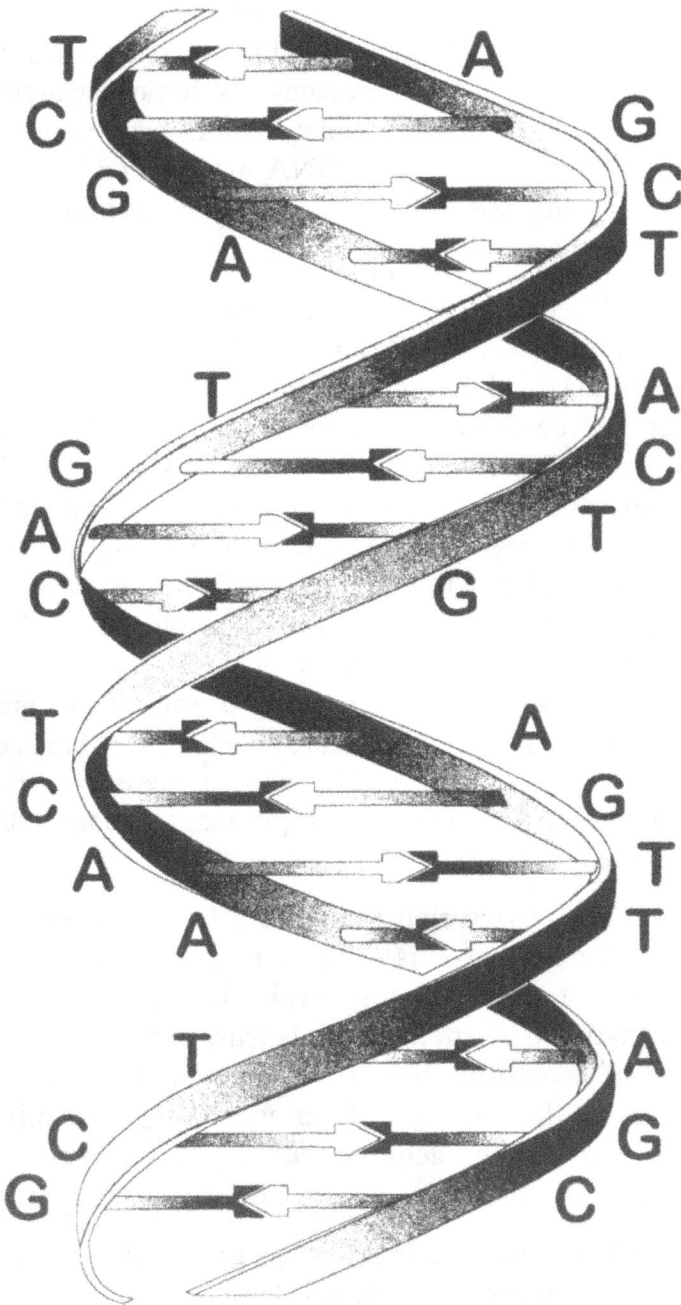

Figure 5. A length of double helical DNA.

the dimensions of the nuclear compartment. In people, the total length of DNA in the 46 chromosomes is 1.8 metres. This is cramped into the nucleus, a spherical structure just six microns across, as extremely thin intertwining threads. As different segments of DNA are needed at different times during the life of the cell, they unfold, allowing access to their coded information.

Whether a cell has one chromosome, as in bacteria, or has multiple chromosomes, as in other organisms, its genetic material, or genome, must be replicated exactly, once for each cell division. This is a complicated endeavour. Before replication can occur, the parental DNA strands must be separated. Synthesis of new daughter strands then begins using each of the parental strands as a template (Figure 6). The fidelity of replication is assured by specific base pairing between the old parental strand and the new daughter strand: A with T, and G with C. While this is proceeding, a group of remarkable enzymes move along the new molecule, checking for damages and correcting any mistakes. The whole process is carried out with stunning precision at a rapid rate — about 750 bases are linked up every second.

The observation that DNAs from different species have different amounts of bases raised the possibility that the sequence of bases might be how the genetic information is written. If this is true, then it brings up the issue of explaining how the sequence of bases in DNA represents the sequence of amino acids in a protein. This would lead to the genetic code. For the information in DNA to have any meaning, its code must be deciphered.

The genetic code is read as groups of three bases, or triplets, each representing one amino acid. Each triplet is called a codon. A gene can now be considered as a series of

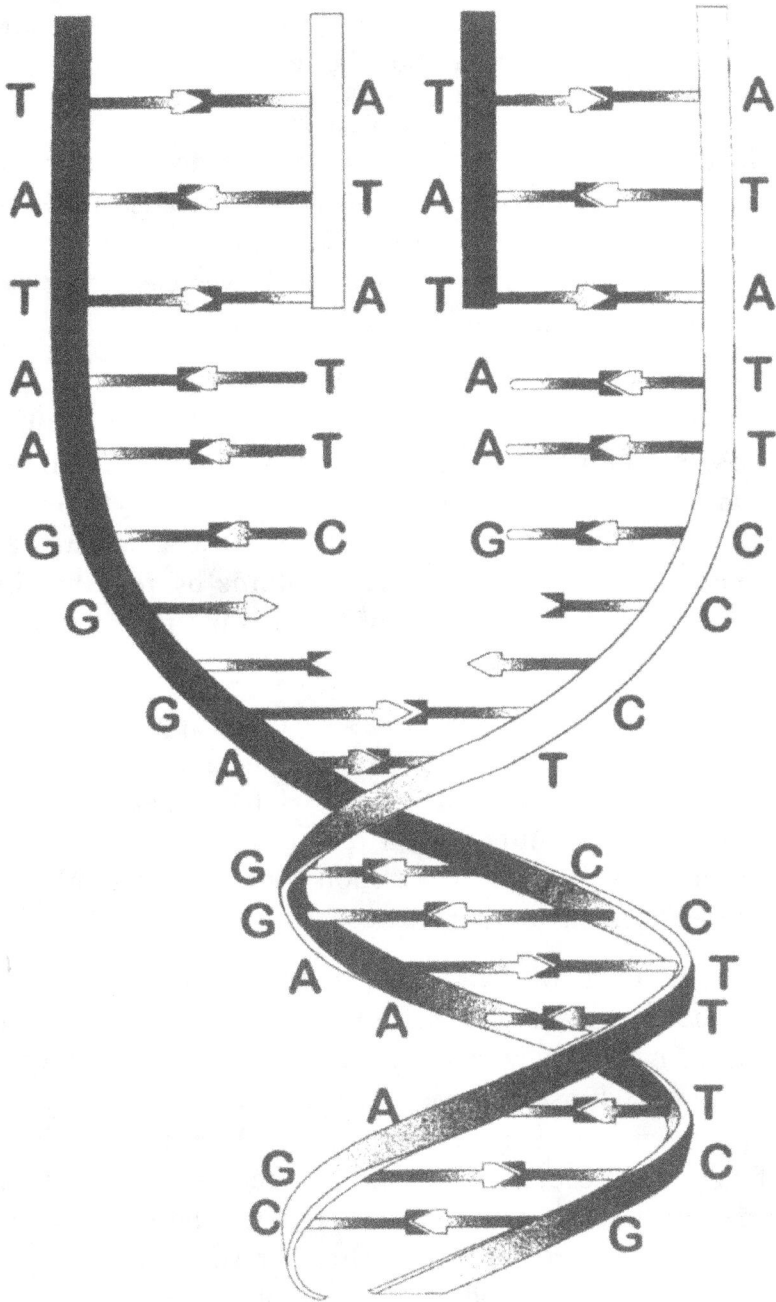

Figure 6. Replication of a DNA molecule, showing strand separation and incorporation of new bases.

codons that is read in tandem from a starting point at one end to a stopping point at the other end.

The breaking of the genetic code revealed that information was stored in the form of codons, but it did not tell how the information was used. Because DNA in the nucleus was physically separated from the site of protein synthesis in the cytoplasm, it was clear that the process was not straightforward.

The information in DNA is, in fact, first copied, or transcribed, onto another molecule, the messenger RNA. As the name suggests, the messenger RNA moves from the nucleus, where it is made, to the cytoplasm, where it functions. Messenger RNA has a fixed purpose in cells: to be translated into proteins by reading the genetic code — each triplet of bases is interpreted as an amino acid.

Any of four possible bases can occupy each of the three positions in the codon, so that there are 64 possible combinations of bases (4 x 4 x 4 = 64). There are only twenty amino acids, much less than the 64 possible codons. This means that each amino acid could be represented by more than one codon.

In 1961, Marshall Nirenberg started the effort to decipher the code, and within five years it was cracked (Table 1). Almost every amino acid is, indeed, represented by a few codons. It turns out that codons representing related amino acids tend to be similar. Often the third base of the codon seems less important because codons with different bases in this position code for the same amino acid. The amino acid glycine, for instance, can be represented by four codons: GGU, GGC, GGA, or GGG; in this instance, the first two bases are identical, while the

Table 1. The Genetic Code

1st position	2nd position				3rd position
	U	C	A	G	
U	Phe	Ser	Tyr	Cys	U
	Phe	Ser	Tyr	Cys	C
	Leu	Ser	STOP	STOP	A
	Leu	Ser	STOP	Trp	G
C	Leu	Pro	His	Arg	U
	Leu	Pro	His	Arg	C
	Leu	Pro	Gln	Arg	A
	Leu	Pro	Gln	Arg	G
A	Ile	Thr	Asn	Ser	U
	Ile	Thr	Asn	Ser	C
	Ile	Thr	Lys	Arg	A
	Met	Thr	Lys	Arg	G
G	Val	Ala	Asp	Gly	U
	Val	Ala	Asp	Gly	C
	Val	Ala	Glu	Gly	A
	Val	Ala	Glu	Gly	G

U* = Uracil   C = Cytosine   A = Adenine   G = Guanine
*For DNA, read T = Thymine, instead of U

Ala   Alanine
Arg   Arginine
Asn   Asparagine
Asp   Asparatic acid
Cys   Cysteine
Gln   Glutamine
Glu   Glutamic acid
Gly   Glycine
His   Histidine
Ile   Isoleucine
Leu   Leucine

Lys   Lysine
Met   Methionine
Phe   Phenylalanine
Pro   Proline
Ser   Serine
Thr   Threonine
Trp   Tryptophan
Tyr   Tyrosine
Val   Valine

STOP means "end chain"

The code is read in order from the first through to the third position.

third is different. Meaning was quickly assigned to 61 of the 64 codons. The three remaining codons — UAA, UAG, and UGA — do not code for amino acids, but signal the stop of protein synthesis; one of these marks the end of every gene.

The genetic code is universal. The identical set of codons is used by bacteria, butterflies, and people. This means that theoretically a messenger RNA from one species would be translated correctly by another species as the codons have the same meaning across these boundaries. This universality of the genetic code indicates that it was established very early in evolution, and passed down through the aeons to all living organisms from that original ancestor.

Codons in a gene determine the alignment of amino acids in a protein. The actual assembly of amino acids into proteins is conducted by the ribosome, a compact little factory, itself made of several proteins attached to a long RNA molecule, called ribosomal RNA. Each ribosome is made of two parts, a large and a small, fitted together to form a tunnel through which the messenger RNA is threaded.

The different messenger RNAs contain the coded information from which the amino acids are lined up. All cells carry out this transaction in the same way. The information coded in DNA is first transcribed onto a messenger RNA, which is then translated into a protein molecule by ribosomes.

As a general rule, the flow of genetic information in a cell is as follows:

DNA — > RNA — > Protein

This relationship between DNA, RNA, and protein led to a profound conclusion, postulated by Crick in the "Central Dogma." He suggested that genetic information travelled one way only, and considered this as absolute with no possibility of reversed flow. DNA issued the blueprints for manufacture of RNAs, which then directed the assembly of proteins. Protein molecules could not work their way back through RNA to DNA. Indeed, there appeared no biological need for the information flow to be reversed.

The doctrine of the Central Dogma endured for some time, but the discovery of viruses was another matter. Viruses, small packets of genes in protective protein shells, invade and hijack the cells of living organisms, forcing them to use their apparatus to make new viruses. While some viruses have DNA as their genomic material, others have RNA genomes. The RNA viruses lead strange life styles, and as more was learned about them, the doctrine of unidirectional flow of genetic information from DNA through RNA to protein had to be revised.

The virus that led to the upending of the Central Dogma was the Rous sarcoma virus, discovered in 1911 by Francis Peyton Rous. He found that a particular type of cancer, called sarcoma, in chickens might be caused by a virus. He demonstrated this by grinding up the cancer, and injecting a filtered portion of it into healthy chickens. These came down with identical cancers. The experiment could be repeated, and in each instance a cancer developed in the chicken injected. Filtering the ground-up cancer removed any cancer cells, but the tiny viruses could pass through the filter. This ruled out the possibility that the cancer in the chickens was due to the unintentional transfer of malignant cells, but was, in fact, due to the virus.

The suggestion that a cancer might be caused by a virus was so controversial at the time that Rous did not use the word to explain what he had found. Some 25 years later, the idea of a virus causing a cancer was more acceptable, and the Rous sarcoma virus was named, even though there was still widespread scepticism surrounding the matter. The Rous sarcoma virus is a relative of a group of viruses that cause leukaemia in mice, which later became known as retroviruses.

Working with the Rous sarcoma virus, Howard Temin made an important observation in 1960. Usually, when a virus infected a healthy cell, new viruses were made, and the cell died. However, when chicken cells were infected with the Rous sarcoma virus, they not only produced new viruses, but they survived and continued to divide, eventually becoming cancerous. A possible explanation for this was the virus brought into the cell new genetic material that drastically altered its behaviour.

A year later, the discovery that the genome of the Rous sarcoma virus was RNA, and not DNA, made things even more complicated. To overcome the obvious problems posed by this, Temin proposed that the Rous sarcoma virus did not proliferate like other viruses, but its genetic material passed via an intermediate step from RNA to DNA, and then back to RNA. Thus, two steps were involved. First, the infecting virus's RNA was transcribed into a DNA intermediary, called a provirus, and then back into RNA in the progeny virus. With this formulation of the provirus hypothesis, the Central Dogma was revised.

This was a bold suggestion at the time, and final acceptance of the provirus hypothesis came in 1970 with the discovery of an enzyme in the RNA virus, which was capable of transcribing the viral RNA genome into an intermediate double-stranded DNA. The enzyme, aptly called reverse transcriptase, closed the loop. Until it was found, the flow from RNA to DNA, and back again, was conjectural, but now the path for this two-way traffic was defined. All RNA tumour viruses contain this enzyme, and today are referred to as retroviruses. Among its family members is the well known human immunodeficiency virus, the agent of AIDS.

The demonstration that genetic information can flow from RNA to DNA is a pivotal event in modern biology. Because the retrovirus can cause cancer by introducing alien genes, as a provirus, into healthy cells, it has kindled considerable interest in this area, leading ultimately to the discovery of "cancer genes" in human cells.

Equally intriguing is the possibility that cells of animals might use reverse transcriptase or similar enzymes to exploit this transfer of genetic information from RNA to DNA. This was proposed, also by Temin, in the protovirus hypothesis, and some evidence has recently emerged in support of it. The generation of genetic diversity, so commonplace today amongst living things, may have been due, at least in part, to this mechanism. A piece of DNA in one cell is copied into an RNA molecule that is ferried by a virus into a second cell. Here, reverse transcriptase converts it back to DNA, which is then stitched into the cellular chromosome.

Does this actually happen? The potential for conversion of RNA to DNA exists in normal cells, and with this comes the strong possibility that the exchange of genetic material does go on. The DNA of organisms, including humans, seems to have a considerable amount of material that appears to have been created via RNA-to-

DNA mechanisms, followed by integration of the new DNA into the genome. Perhaps, more than 10 percent of the human genome may have originated in this manner.

If true, it means that evolution could have been accelerated partly by a steady traffic between the cellular genome and the outside. Viruses that flit in and out of cells, tugging along pieces of genetic material from one species to another and reshuffling their genetic make-up, may have played a greater part in our genetic destiny than we could have ever guessed.

A system with a close parallel to the protovirus theory is the mobile genetic elements, first described by Barbara McClintock in the 1930s. While performing breeding experiments on corn at Cold Harbor Laboratory, she saw odd patterns in the inheritance of the colour of pigments in kernels, which conventional rules could not explain. These were erratic in that they occurred in a gene for a number of generations; that gene then became stable, whilst another became unstable and susceptible to mutations.

After puzzling over the results, she concluded that a few genes did not have fixed locations on a chromosome. Rather, they seemed to jump from one spot to another between parent and offspring. McClintock suggested that the jumping genes settled near other genes, modifying, changing, their message, and accomplished in one leap a degree of variability that we once believed would take aeons of evolution to achieve.

Up to then, the gene was seen as a stable entity that, barring the occasional mutation, was a comfortable constant in the changing world outside the cell. Not surprisingly then, McClintock's idea of jumping genes languished for decades. But by the 1970s, the movement of pieces of DNA between cells was accepted, and McClintock was given the recognition she justly deserved in 1983 with the award of the Nobel Prize.

The origin of jumping genes is still obscure, but they may be remnants of viruses that have become permanent lodgers in the cell's chromosomes. Jumping genes generally seem to rely on a mechanism similar to the retrovirus for their gymnastics; some make it on their own, but others borrow enzymes from more capable viruses.

However they accomplish their jumps, these genes can exert a powerful influence on the organism bearing them. In some cases, they cause serious diseases. In 1991, Francis Collins and colleagues at the University of Michigan identified a patient with a genetic mutation causing neurofibromatosis, a pre-cancerous condition. A gene that normally regulated cell growth had been rendered inactive by insertion of a common genetic element called *Alu*. Two months later, a team headed by H. Kazazian at the John Hopkins University School of Medicine announced that it caught another gene almost in the act of jumping. While studying a group of haemophiliac patients, they diagnosed the disease in a child, in whom a jumping gene inactivated the gene for one of the blood-clotting proteins. The jumping gene was essentially identical to a gene at a different location in the child's parent. In the parent, the gene was silent, but when it moved to its new location in the child, it knocked out an important gene to cause haemophilia.

Jumping genes are usually passed vertically from parent to offspring, not unlike ordinary genes. However, their ability to take short leaps within cells has prompted speculation about the possibility that, under rare circumstances, they might jump horizontally from one organism to another, perhaps even between different species. In lower organisms, such exchange of genes is known to occur. Bacteria can exchange genes under certain conditions, and transfer of genetic material between bacteria and plants and insects is strongly suspected.

The transfer of genes between higher organisms is as yet unproven. Nonetheless, John McDonald, a molecular geneticist at the University of Georgia, notes that unrelated but cohabiting species often have peculiar genetic similarities. He suggests that genetic information had passed between them, but so far this is speculation. Inasmuch as the mechanism for transfer of genetic material between lower organisms is well recognised, the possibility of a similar occurrence among higher organisms cannot be discounted.

In the 1860s, Mendel proved that traits were passed from one generation to the next through factors, which later became recognised as genes. The discovery of DNA in the 1950s as the essential ingredient of inheritance ignited an intense investigation into genes that is still going on in earnest. At first, the flow of genes through succeeding generations seemed stable, immutable. This is exactly what we would have predicted. Genes are carefully copied and faithfully passed on. Order is maintained and ensured through the rigid laws that govern genetics. We roll ourselves through generation after generation by passing down our genes.

We have seen an upheaval in this thinking in the past decade. What we are now learning is that the genome is not reliably static, but is constantly changing. This has some deep implications. What are the consequences, good or bad, of this genetic agility? And in the face of such fluidity, how can any order be achieved? In the chapters to follow, we shall explore these issues.

CHAPTER 2

THE GARDEN OF EDEN

A remarkable observation of modern biology is how similar all living things are once you scratch below the surface. The deeper we look at the molecular level, the more alike everything appears to be. A reasonable explanation of this inner uniformity, more astonishing than the outer diversity, is of course that all living things share a common origin, descendants of a single ancestor or primogenitor. This is certainly a valid assumption, and provides a convenient point of departure to find our common mother.

What is certain is that the Earth was quite a different place some 4.7 billion years ago when it condensed from a whirling cloud of gas and interstellar dust. It was a violent environment, passing through cycles of heating and cooling, and undergoing great upheavals that continually reshaped the landscape. Even so, it did not take long for life to gain a foothold. Paleobiologists, who study ancient rocks and fossil sediments, found that the first fossil cells, or microfossils, evolved when the primitive Earth was just one billion years old.

What happened in those billion years before life was conceived, before the first beat of a heart? What molecules arose at this stage to made life possible? To frame the question another way, we can ask what is the simplest molecule or collection of molecules that can straddle the transition between the inanimate world and the earliest stirrings of life.

Looking back, we would reasonably expect that the prerequisite for any living system must be its capacity to propagate, allowing it to make similar, if not identical, copies of itself. It is a modest claim for membership in the club of life, but an essential one. Without the capacity to replicate, molecules, no matter how suited to form the underpinning of life, would go nowhere.

If we accept this premise, then we can scrutinise today's cells for footprints, however faint, on this ancient path. We can do so because evolution is a conservative process; the old is not discarded, but retained and modified for new functions, leaving telltale relics of itself. Careful reading of these allows us to piece together the nature of the primogenitor.

In the molecular Garden of Eden that was the early Earth in its first billion years, Nature conducted many experiments. We shall probably never know the details of all of these, but we can with some confidence retrace the steps of the few that eventually led to the origin of life. We can speculate that the universal presence of DNA in all contemporary cells, and many viruses, makes it an easy front runner as a candidate for the ancestral molecule of life.

In its double-helical form, DNA seems perfectly up to the job. Its two strands can serve as templates, which are copied in the creation of new strands. This ensures that the primordial molecule leaves copies of itself, an important requirement if the molecule is to maintain its identity as it goes forth and multiplies.

We immediately face a problem. It is difficult to understand how a DNA molecule, or anything like it, could replicate on its own. Conceptually, replication of DNA may be simple; the copying of each of the two strands occurs by the pairing of the bases — A to T, and G to C. The actual process is, however, more involved. The strands of the DNA double helix are twisted around each other, and before they can be copied, they must first be unwound. After this unwinding, the two strands are unzipped, and only then can new bases be brought in to form a new strand. All of this requires catalysis by proteins.

Nowadays, DNA is synthesised only with the help of proteins; without them, nothing would happen, or would proceed at such a snail's pace as to be next to nothing. This brings up the troubling question of where did the proteins come from. It seems the classic situation of the "chicken and the egg" or, in the case of science, it's back to the drawing board.

DNA contains the information for the making of protein molecules, while proteins catalyse the replication of DNA. It is extremely improbable that DNA and proteins, both complex substances, arose at the same time in the same place. Today, we cannot have one without the other, and this causes an uncomfortable dilemma.

A way out of this impasse was suggested in the late 1960s by Leslie Orgel at the Salk Institute for Biological

Sciences in San Diego. He proposed that RNA, and not DNA, might be the first molecule of life. In this scenario, RNA catalysed all the reactions necessary for the common ancestral molecule to survive and replicate, and it did this without proteins. Later, however, RNA would develop the ability to join amino acids together into proteins. In this view, two properties of RNA, not immediately evident today, were needed — a capacity to replicate without proteins and an ability to catalyse protein synthesis.

Proof for Orgel's idea came with a startling discovery in 1981 by Thomas Cech of the University of Colorado at Boulder. Working with a large, unicellular aquatic creature, *Tetrahymena*, he found that its ribosomal RNA also served as a biological catalyst. Of relevant importance, it did so without the help of a protein. Until this discovery, only proteins were thought to be catalysts in living systems. However, the demonstration that an RNA molecule could be a catalyst immediately lent credence to the proposal that one of its kind might, indeed, be the very first "living" molecule. The first ribozyme, as these RNA catalysts are now called, could do little more than cut and paste pieces of RNA together. Nonetheless, the recognition of the catalytic capacity of RNA was significant, and shed unexpected light on a central problem of the origin of life.

Following the description of the first RNA catalyst in 1981, a second was discovered two years later. Since then, many other instances of catalytic RNAs have been found in contemporary cells, molecular fossils of earlier times. The catalytic property of RNA is, therefore, not an isolated quirk in *Tetrahymena*, but a widespread occurrence among living organisms.

The RNA molecule was also capable of carrying genetic information, coded in the sequence of its bases, and this made the concept of it as the originator of a living system more plausible. So far, however, no natural RNA molecule has been found that directs its own replication. Recently, Cech and Jack Szostak of the Massachusetts General Hospital have succeeded in modifying naturally occurring ribozymes so that they could string together few RNA bases. It is indirect evidence, but it, nevertheless, supports the notion that ancient RNA molecules might have been more accomplished than their descendants around today.

A new view is taking shape. In the chemical cauldron of the early planet, simple elements interacted to form more complex compounds similar to those now found in living things. The atmosphere at the time contained very little oxygen, but was rich in hydrogen and other gases like methane and ammonia. In 1953, Stanley Miller and Harold Urey at the University of Chicago conducted the first experiment to find out what sort of chemical reactions would take place in such an environment. In a self-contained apparatus, they created an atmosphere of hydrogen, methane, water, and ammonia over an "ocean" of water. They exposed the gases to electrical discharges to simulate "lightning." The experiment yielded a number of organic compounds, including several amino acids. Later, other studies showed that the components of RNA and DNA could also arise under similar conditions.

We can look back into the distant past and put the bits together. It seems certain that all the necessary chemical building blocks were at hand during the first billion years, distilled from the intense reactions fuelled by the sun, ionising radiation from outer space, and volcanic heat. The bases joined together, forming short, rudimentary chains of RNA. Over time, these gradually became longer and more elaborate, while mutations added a richness to their variety. Many, perhaps almost all, of these chains might have been useless, but an occasional RNA molecule arose that had the right configuration capable of engineering its own replication or that of other RNA molecules. These left behind copies of themselves, slowly spreading across the surface of the Earth. The first pulse of life, and a humble beginning, indeed.

The catalytic RNA in *Tetrahymena* is made of a long string of 413 bases that folds into a compact molecule. Other catalytic RNA molecules also tend to be as large and complex, and this brings up a worrying issue. It could be argued that to construct such a large, functional molecule by the random assembly of bases might be unrealistic. However, every catalyst has a crucial "active" site, which is usually much smaller than the whole molecule. It is through the special spatial configuration of this much smaller area that the catalyst exerts its effect; the rest of the molecule serves merely to align it properly, allowing it to dock with the molecules it eventually reshapes. Conceivably, then, the first RNA catalyst might have been considerably smaller than those found today, but, nevertheless, capable of useful function.

Cogent evidence for this comes from some viruses that infect plants. These have relatively small, single-stranded RNA genomes, short segments of which are active catalysts. From this, we can surmise that the catalytic potential of RNA does not require it to be long; only about 50 bases are sufficient. The scale of this might have been achievable in early biological systems.

Early RNA could store information with a simple system of four bases, arranged like beads along a string. The memory of this information was passed on to new molecules that were created in its likeness. The strength of this was that the accomplishments of one generation, secured in the linear language of its bases, were passed on to the next. As the original, crude RNA became progressively refined over time, all its information, old and new, was kept.

The emergence of RNA was a watershed event in the development of life, but progress was stalled in the RNA world because of the limitations of RNA molecules as catalysts. For efficient catalysis, the RNA strings must fold upon themselves to form exact three-dimensional surfaces. Unlike proteins, the RNA molecules have a restricted repertoire, and this would eventually lead to a plateau in the progressive enrichment of evolution.

Inherent in its make-up, the RNA molecule already stored genetic information, which was passed on from one generation to the next. What was needed was a means to use the RNA information to make proteins to run the show. The design of a mechanism for the production of proteins, under the direction of RNA, introduced a new stage of advancement, the ribonucleoprotein world or RNP world.

This was a quantum leap in sophistication of living systems. The dual role of RNA to store information and to execute its own message would now be split. We begin to see the emergence of the legislative — RNA — and the executive — protein — branches in the RNP world.

Today, the RNA message is decoded, or translated, by the ribosome, an elaborate molecular machine made of a long RNA molecule and 50 different proteins. It is a complex apparatus, not one that arose in one big swoop, but over perhaps millions of years as Nature tinkered one step at a time. Without doubt, the original ribosomal machinery for translation of RNA into protein was crude, but became increasingly more efficient as time went by.

Even today, we can still find evidence in the ribosome for the existence of the RNA world. The ribosome travels along the strand of messenger RNA, reading its message, which it uses to link one amino acid to the next by forming a chemical bond between them. Henry Noller of the University of California at Santa Cruz found that it is probably the ribosomal RNA, not its proteins, that catalyses the formation of the chemical bonds between the amino acids. As RNA preceded the arrival of proteins, this essential function of the ribosome was initially carried out by RNA, and it has remained so ever since.

Francis Crick had a crucial insight into how specific amino acids were brought together and linked to form a protein molecule, and how the message in RNA ensured that the pattern was repeated time and again with reasonable accuracy. In 1955, he predicted that there were special molecules that acted as a go-between, recognising the RNA base sequence on one hand and the corresponding amino acid on the other. Their role was to bring to the RNA the correct amino acids for which its bases coded. Crick's suggestion was confirmed with the discovery of molecules that served as adaptors between RNA and amino acids. The adaptors were other molecules of RNA, appropriately called transfer RNAs. They provided that important means to translate a triplet codon in RNA into an amino acid in protein.

Each transfer RNA molecule has two important sites. The first sticks to a specific amino acid; the second is a group of three bases, referred to as an anticodon, that is complementary to a triplet codon in RNA. The docking of the anticodon on transfer RNA to the appropriate codon on messenger RNA ensures that it delivers its amino acid at the right spot. The genetic message is, thus, correctly translated into a protein.

Transfer RNAs from all species look very much alike. They are long strings of about 74 to 95 bases, which fold upon themselves, bringing short segments together that form cross-links between complementary bases: A to U, and C to G. They take on a cloverleaf pattern with three loops and a free arm (Figure 7). The free arm attaches to the amino acid; at the opposite end is the anticodon loop with its three bases that match those on the codon in the RNA. There are twenty different transfer RNAs, one for each of the amino acids.

The transfer RNA pairs up with its amino acid, and they are swept onto the ribosome. If the anticodon matches the codon, the transfer RNA is welcomed into the ribosome, where it releases its amino acid; if the fit is loose, the pair caroms away and tries again until there is a proper match. The transfer RNA, thus, fulfils two purposes; it seeks out its specified amino acid, and it brings it to the ribosome where it is added to the nascent protein when the correct codon is present.

In modern cells, the translation of the RNA message occurs efficiently because of a second molecular adaptor that brings the transfer RNA and its amino acid together. This speeds up the interaction between the two. In the normal course of events, the transfer RNA and amino acid float around at random. Only when the transfer RNA drifts by its specified amino acid, do the two lock together. This process of sorting and selecting can be slow and inefficient. The second adaptor hunts down the amino acid and the transfer RNA and brings them together, greatly enhancing the likelihood of the two pairing up.

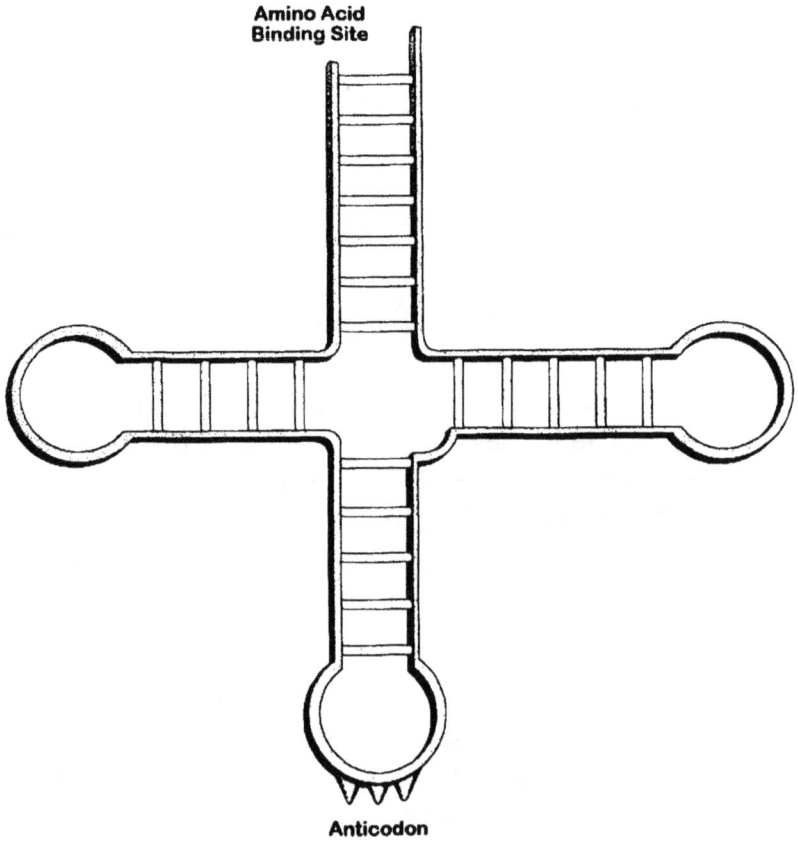

Figure 7. Transfer RNA drawn in its clover leaf form.

Unlike transfer RNA, these second adaptors are all proteins in today's cells. A most unexpected finding is that the adaptors are quite different from one another — some are small, others large; some are made of single protein chains, others of two intertwined chains. At a more basic level, the amino acid make-up of the different adaptors displays no similarity; there are no identical stretches of amino acids, however short, in different adaptors that might betray a common kinship. Each protein is quite different from all others, pointing strongly to a different original source for each.

The marked diversity in the second adaptors flies in the face of a recurring theme in evolution: a common ancestor which slowly changes, one step at a time, over evolutionary time span. It is easier for a species to modify some of its proteins and put them to new uses than it is to assemble new molecules *de novo* every time. In this traditional mode of thinking, we should expect that the second adaptors descended from a common ancestral protein, but had undergone changes over time that allowed them to catalyse the reactions between the different transfer RNAs and their amino acids. In this scenario, we would expect the different second adaptors to show some degree of structural similarity.

The striking differences among the second adaptors were a puzzle at first, but there might be a plausible explanation. At the beginning of the RNP world, all adaptors were RNA molecules, the only species of biological molecules around at the time. As the RNP world evolved, however, these RNA adaptors were gradually replaced by proteins, which are still around today. It is reasonable to expect that the replacement of each of the older RNA adaptors by the younger, more efficient protein adaptors took place independently. This would account for the diversity of the protein molecules that took over as adaptors. Here, the crucial element is the proper spatial and three-dimensional configuration of key points in the molecule, not its overall size or shape.

Let us now return to the ribosomes, the most important invention of the RNP world. The adaptors brought the amino acids to the RNA that determined the order in which they were arranged in the protein. The first primitive ribosomes arose to accelerate the linking together of the amino acids, two at a time, by providing a workbench to align the amino acid-transfer RNA complex next to each other. As they became more adept, longer and longer strings of amino acids were constructed, which folded upon themselves to form protein molecules. The way in which a string of amino acids folded to form a compact molecule of protein was dictated largely by the chemical composition of the different amino acids and in what order they were arranged; it occurred easily and effortlessly, requiring no further input from RNA.

As the more versatile protein molecules slowly took over as enzymes from RNA, biochemical reactions among a wide range of compounds became possible. A framework for a workable "living" system was laid down. RNA coded the genetic information, passing it from one generation to the next; proteins read this information and carried out its instructions.

Eventually, living systems became so complex that RNA was no longer a suitable genetic material. The main reason was that RNA could be damaged, especially under the extreme conditions that prevailed in the early Earth, causing a change in its make-up, a change in its stored information. This very feature might have made RNA the ideal molecule to get the whole process started. Rather than a hindrance, the susceptibility to change would have given it the leeway needed to explore all possible avenues of biochemical reactions in quick succession. The reactions that proceeded rapidly had a head start and would become dominant.

Thanks to the new proteins, biochemical reactions sped up and became more connected, paving the way for an integrated metabolism. Under these conditions, the instability of RNA became a liability. Once a species had taken hold, wide swings in its genetic information could not be tolerated if it was to survive. Success depended on the reliable production of proteins, and the information must, therefore, be passed down more faithfully and reliably. Major changes in the genetic material would leave species stranded in an evolutionary *cul-de-sac*.

One way to overcome the problem of RNA's susceptibility to change was to develop double-stranded molecules with complementary strands of RNA. This allowed any dire changes or errors in one strand to be corrected using the other strand as a reference. Even so, the emergence of a duplex RNA would not be expected to be an ideal solution. In fast growing organisms, too much time and energy would have to be expended in repairing damages to the RNA strands.

The transition from RNA to DNA solved this problem, since DNA is more stable than RNA. Like the shift from the RNA world to the RNP world, this third step in the evolution of the gene was a major one; it would permit the development of more complex living systems, and consolidated the gains achieved in the RNP world.

Important as it was, it was but a gentle turn in the path from RNA to DNA. All organisms today make the basic units of DNA from those of RNA in a single biochemical step, and it is, therefore, reasonable to assume that early living systems had at hand the ingredients for the transition to the DNA world. Further, if RNA had already adopted a double-stranded form in an attempt to become more stable, the enzymes that catalysed its replication could be press-ganged into service for copying the original RNA into DNA.

This view of events suggests that it was not a difficult step to go from RNA to DNA. Indeed, even today some viruses, like the retroviruses, slide from an RNA genome to a DNA provirus with ease at different stages of their life cycle. While the transition from RNA to DNA might not have been technically overpowering to achieve, its implication for the whole of evolution of life is boundless.

It is salient to point out here that the RNA-protein transactions worked out in the RNA and RNP worlds remain largely intact in the DNA world. The ascension of DNA has not subverted the biochemical processes already in place. In fact, in contemporary cells, DNA is first transcribed to RNA, around which the workaday details of their metabolism revolve.

Three of the four bases are identical in both RNA and DNA — adenine, cytosine, and guanine. The fourth base is thymine in DNA and uracil in RNA. These are very similar chemically, and, in fact, uracil is converted easily into thymine in a single biochemical step. The substitution of uracil in RNA for thymine in DNA affects the chemical mechanisms, but not the meaning of the code.

It is most remarkable that the enzyme for converting uracil to thymine is identical in viruses, bacteria, yeasts, and vertebrates, in all of life. This strongly suggests that the mechanism for the conversion occurred only once, and very early in the DNA world. It has been conserved, unaltered, in all cells and their descendants ever since.

We can speculate why the substitution of thymine in DNA for uracil in RNA might have occurred. The base cytosine can occasionally be degraded to uracil under natural conditions. This mutation occurs at a low rate, but over the long evolutionary time span, the harmful effects of the change are sure to add up. To prevent this, an enzyme searches for and removes any uracil from DNA, replacing it with the original cytosine. Had uracil remained a base in DNA, the enzyme would not distinguish between an original uracil base and one formed through mutation of a cytosine base; it would remove both, replacing them with cytosine. This would inevitably lead to considerable alterations in the bases of DNA over time. The change from uracil to thymine in DNA was, therefore, important to avoid the wholesale removal of all uracil bases from DNA.

Early in the youthful Earth with its violent and capricious environment, simple elements came together to form more complex compounds. Bases joined together to form short chains of primitive RNA molecules that made crude copies of themselves. At first, every catalytic reaction had to be carried out by RNA itself, but as the RNP world evolved, proteins became more important and gradually took over the role of catalysis. Some RNA enzymes eventually evolved into RNP machines, such as ribosomes, which are still around to this day. Other RNA enzymes lost their RNA component and were replaced by proteins. The transition to proteins enabled living systems to become far more elaborate, for they were more versatile as catalysts than RNA. Once a level of complexity was attained, large changes in the information coded by RNA could not be tolerated, and the transition from RNA to DNA as the keeper of our genetic history took place. Even so, the accomplishments of the earlier RNA world were kept intact, and traces of RNA's former grandeur still abound.

They are with us to this day.

CHAPTER 3

IN SEARCH OF EVE

For about the three centuries we have been doing science, we have had a hankering to classify things, labelling and putting each in a pigeon-hole. It is a tidy way to keep everything sorted. This tradition followed that of the arts and philosophy in the eighteenth century when the overriding concern was with order. There was the strong sense that behind this order there were natural laws for everything.

The naming of living things and arranging them in groups reflected the belief that there were patterns in Nature, into which we could get a glimpse by a systematic method of classification. The most prominent figure in this endeavour was Carolus Linnaeus, a Swedish botanist. He began his system of classification as early as 1730, and by 1749 proposed his binomial schema for naming organisms, which is still in use today. In this, each organism is given two Latin names, one for the genus and the other for the species. Humans, for instance, belong to the genus *Homo* and the species *sapiens*; our name, *Homo sapiens*. The garden pea is *Pisum sativum*; the common gut bacterium is *Escherichia coli*.

The first and major line that Linnaeus drew was the distinction between plants and animals. An organism was either one or the other; it was a simple, unambiguous scheme. It seemed, at first glance, an obvious way of looking at living organisms. We can easily recognise plants; they make their own food from carbon dioxide and chemicals around with a green substance, chlorophyll, that allows them to use the sun's energy. Animals, on the other hand, are unable to do so; they feed on plants or other animals.

Linnaeus's classification endured for well over a century. But the discovery of bacteria proved a troubling matter. Linnaeus had never heard of them, but, in fact, they swarmed over the entire Earth, where they made a good living for nearly four-fifths of its history. They may seem uninspiring and uninteresting, but they could not just be swept under the rug and ignored.

So, what about bacteria? They are clearly different from animals and plants. Many do not have chlorophyll, and thus, cannot be called plants. Moreover, they make their food directly from simple chemicals, and are, therefore, unlike animals.

The difficult issue about bacteria was resolved in 1866 by a German biologist, Ernst Haeckel, who proposed the idea for a third kingdom. All the single-celled organisms, like bacteria, could be lumped together into a new kingdom, which he called monera, meaning neither plant nor animal.

Haeckel's classification worked for a while, but it was given the heave-ho when biologists peered down their microscopes at the myriad organisms that made up monera. The convenient inclusion of all of them into one kingdom just because they were single-celled would not do. Among the single-celled organisms were some that were truly more like plant cells, and others that were more like animal cells. The principal distinction between them and bacteria was the presence of a cell nucleus.

Every plant or animal cell has a distinct nucleus that houses its chromosomes. The bacterial cell, on the other hand, does not have a well-defined nucleus, and its single chromosome lies free in the cell. On this basis, bacteria are different from other single-celled plants and animals, and any attempt to squeeze them together into a single kingdom leads to an awkward dilemma.

Despite the superficially bland nature of bacteria, they could not be easily dismissed. They occupied and spread themselves across the Earth, where they were the only inhabitants for a very long time. Far from an aimless existence, a prelude of things to come, these simple organisms engaged in learning everything for surviving. Biological novelties emerged right and left, some workable, others, perhaps most, not. And, around a billion years ago, after all the skills for survival had been mastered, a great evolutionary leap occurred: the invention of the nucleated cell.

For more than half a century, biologists have recognised this important distinction between living things. The simpler systems of Linnaeus and Haeckel had to be scrubbed. The fundamental difference between organisms without a nucleus and those with a nucleus demanded attention, and the obvious and easy division of all living things into animals and plants, or the accommodation of the single-celled organisms into a third kingdom gave way to a new scheme. Two groups have been named: prokaryotes — organisms without a true nucleus — and eukaryotes — organisms with a well-defined, discrete cell nucleus. With this distinction, everything appeared tidy once again.

Up until the 1950s, most of our ideas of evolutionary theory were based on fossil records left behind when living matter turned into stone. In the late 1950s, paleobiologists found that structures resembling modern algae and bacteria — microfossils — could be discerned by microscopic examination of extremely thin slices of certain ancient rocks, stromatolites, from Australia and South Africa. The stromatolites are complex mounds of laminated rocks in which the different layers have been laid down by colonies of various species of prokaryotes. The ancient rocks built up slowly in shallow seas when floating mats of algae and bacteria multiplied and gave rise to one layer of progeny after another.

One trouble with microfossils is that in the process of fossilisation only the hard outer walls of the cells have been preserved. The delicate, soft organic matter was destroyed in the repetitive cycles of heating and cooling. Thus, they tell us little about how the first living thing was put together.

We can reconstruct the events of deep antiquity in another way. Surprising and improbable as it may seem, it is to look inside contemporary cells for old relics. The trick is to find structures shared by such diverse organisms that they must have been present in the oldest of living things and, perhaps, the primogenitor of all of Earth's creatures.

In approaching the problem in this manner, we must resist the temptation to assume that evolution always improves genes. Some genes may need no improvement at all. Several biochemical processes, like DNA replication, transcription, and protein synthesis, are so basic to all cells that they are remarkably similar across all the boundaries of different species. The more critical a function is to a cell, the more highly conserved it is. Hence, it is not uncommon to find some vital proteins have remained largely unchanged in widely different species over the entire course of evolution.

Carl Woese, a bacteriologist at the University of Illinois at Urbana, set about to scrutinise the simplest of life forms, bacteria, for molecular "fossils" that might provide clues about the nature of the earliest cell. It was a novel venture, and one that was full of surprises.

All living organisms share a common feature: a genetic repository which stores information that is passed on to its progeny, and which contains the instructions for everything that a cell does. The genetic material is faithfully copied and passed on; it is an immortal connection threading its way from the earliest ancestors to contemporary organisms, through every cell over billions of years. When we look inside a modern bacterial cell at its genes, we are also getting an insider's view of its remote ancestor.

Some proteins are so essential for life that they must be present in all cells. Without them, life would be unsustainable, and a cell with this misfortune perishes quickly. The genes that code for these proteins must, therefore, be present in the earliest cells. On the other hand, genes that code for proteins which are not so critical for survival display a wider flexibility in their make-up. Minor variations can be tolerated. Indeed, it is this very genetic variability that fuels the evolutionary engine; without it, evolution would have been stopped in its track with no new, random forms to exploit the selective pressures of Nature.

Both of these types of genes are needed to reconstruct the molecular evolution of bacteria. The conserved genes, handed down from the ancestor, are present in all cells, whilst the more diverse are a more recent invention. By knowing what is old and what is new, molecular evolutionists can recreate the family tree of living creatures.

When Woese began to trace the evolutionary journey of bacteria, the technology was not in place to manipulate DNA with ease and rapidity. He chose instead a surrogate, the RNA molecules that make up the ribosomes, themselves relics of the RNP world, which assist in the assembly of proteins from amino acids. Each ribosome has its own RNA, known as ribosomal RNA. There are two major types of ribosomal RNA, a short form of about 1,500 bases and a longer form of about 3,000 bases. Using the smaller subunit of ribosomal RNA, which he could snip, sort, and analyse, Woese hoped to find a spectrum of base patterns that changed at different rates over time. The ancient parts would change the least, and the more modern parts would change more rapidly. He catalogued the fragments of the ribosomal RNA subunit, and began to trace the relationship between various species of bacteria through the similarities and differences in their RNA fragments.

Woese compared the bacterial ribosomal RNA fragments with similar fragments from some larger, more complex, single-celled organisms. He expected that the bacterial fragments would fall neatly into one pile and those from the larger organisms into another. This is exactly what we would have guessed; the prokaryotic bacteria are different from eukaryotic cells, and this would be reflected in their RNAs.

But it did not turn out this way. It is an example of how difficult it is to predict where the next significant piece of information in science will come from. We cannot call the shots in advance. Science moves forward by producing a surprise, always unpredictable. And Woese's surprise would eventually shed light on the most elusive enigma of biology: the real parents of us all.

The ribosomal RNAs of the various bacteria were quite different from the larger, eukaryotic organisms. No surprise here. But some of the bacterial RNAs were equally different from each other. The unavoidable conclusion is that bacteria, the most primitive living organisms, are not all the same, but they come in two very different types.

And so, again, the neat division of living things into two groups — prokaryotes and eukaryotes — accepted for more than half a century had to be revised once more. Woese's work suggested that at the bottom of the tree of life there are not two, but three branches.

The groups of bacteria that caused this upheaval have themselves been a curiosity in the bacterial world. They do not seem to fit together, but have, nonetheless, been put into one basket, because of one peculiar feature — they are anaerobes. In a way, this should not be entirely surprising, as these ancient organisms must have arisen at a time when the Earth's atmosphere contained no oxygen, or at most only tiny traces. Oxygen kills these bacteria. Today, they still inhabit the Earth, but flourish only where there is no oxygen. They abound deep in the mud of swamps or in the shelter of other bacteria that use up any oxygen in the vicinity.

One group is the methanogens, or methane producers, that make a good living in the mud at the bottom of stagnant bogs, in sewage-disposal plants, and in hot vents at the bottom of the sea. They go about their business of living using water and carbon dioxide for food and energy, and in the process give off a gas, methane or marsh gas, that bubbles up from the bottom of swamps.

An even weirder group is the halophiles which live in water that is saturated with salt. They survive in environments, like the Dead Sea and the Great Salt Lake, that will kill almost every other living thing. A third group, the thermoacidophiles, may look like conventional bacteria, but live and reproduce themselves enthusiastically in extremely hot acid waters of hot springs, where the temperatures are close to the boiling point. Chill them down to room temperature, and they die. Another group is the thermoplasms, delicate mycoplasms without a rigid cell wall, that also share the ability to withstand a high degree of heat, making the smouldering piles of coal tailings their home.

At a superficial level, there seems to be no common bond among these unusual bacteria. But look deeper and a family likeness emerges. They are related through their RNA. The RNAs of these bacteria are different from the other common bacteria, so different in fact that they are grouped into a new kingdom, called archaebacteria, a name that denotes their antiquity, for that is exactly what they are, ancient life forms. The other common bacteria are called eubacteria.

The ability of archaebacteria to live in these hostile environments makes them attractive claimants to the title of the progenitor of all cells. There is firm geological evidence that four billion years ago the Earth was a completely different place from what we now live in. It was a stiflingly hot place with no free oxygen in its atmosphere. Any life form arising then must be able to tolerate those prevailing conditions. And the archaebacteria appear quite equipped to do so.

What can we make of these archaebacteria with their bizarre lifestyles? Capable of life under conditions of high temperature and no oxygen, they arose a long, long time ago. They are still around today, but confined to a few hot spots. One possibility is that their descendants have learned to survive in the changing environment and now dominate the cooler earth with its oxygenated atmosphere. But it is not so simple. The RNA analysis cannot be reconciled with this scenario. The RNA difference between the two types of bacteria is enough to cast doubt that eubacteria have descended from archaebacteria.

If we accept the RNA evidence, then we are left with the question of where did the eubacteria come from. Whatever their origin, they have been able to survive the first appearance of oxygen in the atmosphere, perhaps, some three billion years ago when the first organisms capable of photosynthesis appeared. These used nothing more than sunlight, water, and carbon dioxide to make their food, and in the process gradually dumped oxygen into the air. They didn't want it for their own use; it was a by-product of their solar-powered metabolism.

The atmospheric oxygen level increased steadily, reaching one percent about two billion years ago. When substantial quantities of oxygen had built up in the atmosphere, the bacteria around developed the biochemical mechanisms to use it for their energy needs, and about a billion years ago, a new species, the aerobes, emerged. By using oxygen to burn their food, the aerobes were able to extract more energy and were, therefore, more efficient at this than the anaerobes.

Today, the aerobes are all around us, in great abundance in every conceivable niche. They flourish in a world with about 20 percent oxygen in the atmosphere. We oxygen breathers tend to take a relaxed attitude to the oxygen in the air today, but we cannot live without it. Without the invention of photosynthesis by bacteria three billion years ago, we would not be here.

It is tempting to regard the eubacteria as a more recent invention, a group that arose when there was a whiff of oxygen around in the air. In fact, however, the aerobes, those oxygen-loving bacteria that we are more familiar with, are not the only eubacteria around. A lot of eubacteria are anaerobes, which live in areas where oxygen does not penetrate. Just a few centimetres below the surface of the ground, for example, there are swarms of anaerobic eubacteria, endlessly busy with their own affairs, making a living without oxygen.

It now appears that the anaerobic eubacteria are as old as their archaebacterial cousins. As the conditions of the early Earth changed, the eubacteria were able to adapt more quickly than their stodgy cousins, which were left behind to eke out a living in the shrinking environments where they were more comfortable. The eubacteria and archaebacteria are different today, and they have been different for many millions of years.

Until these discoveries, the accepted view was that bacteria belonged to a large, single group. They were the sole occupants of the Earth for most of its life span, during which they began the long, slow process of learning how to get on in the changing Earth before the great evolutionary jump — the emergence of eukaryotes. The idea that eukaryotes are direct descendants of prokaryotes is not the full story.

The reason is that there are two distinct groups of bacteria, which have followed different evolutionary paths. The more venturesome eubacteria adapted to the changing modern world, whilst their first cousins, the archaebacteria, remained anchored in the ancient world of their birth.

The evidence points compellingly to a single ancestor from which a three-way split into three kingdoms — archaebacteria, eubacteria, and eukaryotes — occurred early. Before this, there was, in all likelihood, a furious jostling as cells vied for a foothold in the youthful planet. Once established, however, the three kingdoms dominated, and over time their progeny mastered the skills of life.

It was not immediately apparent where the branching of the three kingdoms from the main ancestral trunk occurred. Did the two bacterial kingdoms — eubacteria and archaebacteria — split off close to each other? And what of the eukaryotes? The answer is coming in and, again, it is a source of some surprise.

Contrary to expectations, several archaebacterial proteins used for gene transcription resemble those of eukaryotes much more than those of eubacteria. A close similarity has also been described in crucial regulatory genes in eukaryotes and archaebacteria, but not in eubacteria. These observations make it likely that archaebacteria, oddities in today's world, may be more closely related to the all important eukaryotes, like us, than they are to the common eubacteria. This is certainly not what we would have guessed.

The similarities of genes and proteins for DNA transcription in archaebacteria and eukaryotes force us to take a new look at the relationship among the three kingdoms, however much anguish it may cause. The archaebacteria and eukaryotes share an ancestral lineage, a common branch that split off from the eubacteria.

We can now sketch a new tree of life (Figure 8). The diagram depicts the relationship among the three kingdoms. Our close neighbours on the evolutionary tree are those strange archaebacteria. They diverge less from the ancestor than eukaryotes, which have surged forward in leaps and bounds with more options for diversity. What is surprising about the tree is how relatively recent the split is between plants and animals, the division that was so prominent in Linnaeus's attempt in the eighteenth century to classify living things.

We can now try to speculate on our ancestor, that very first successful cell from which we, and all other living things, are the lineal descendants. It was, beyond argument, just that: a single cell, capable of replicating itself.

All living things, from bacteria through fruit flies to people, carry out the business of life in much the same way; they all use the same biological manual. The instructions are carefully filed in the linear language of the DNA molecule, collected over aeons of evolutionary history. All the cell's activities, from burning carbohydrates for energy to choreographing the complex dance of the chromosomes when the cell divides into two, are dictated by its genes. The genetic material does this through proteins, long strings of amino acids that fold into a bewildering array of compact structures, the tools of the cell's machinery.

If the same connection between DNA and proteins is invariably observed in all modern cells, is it not

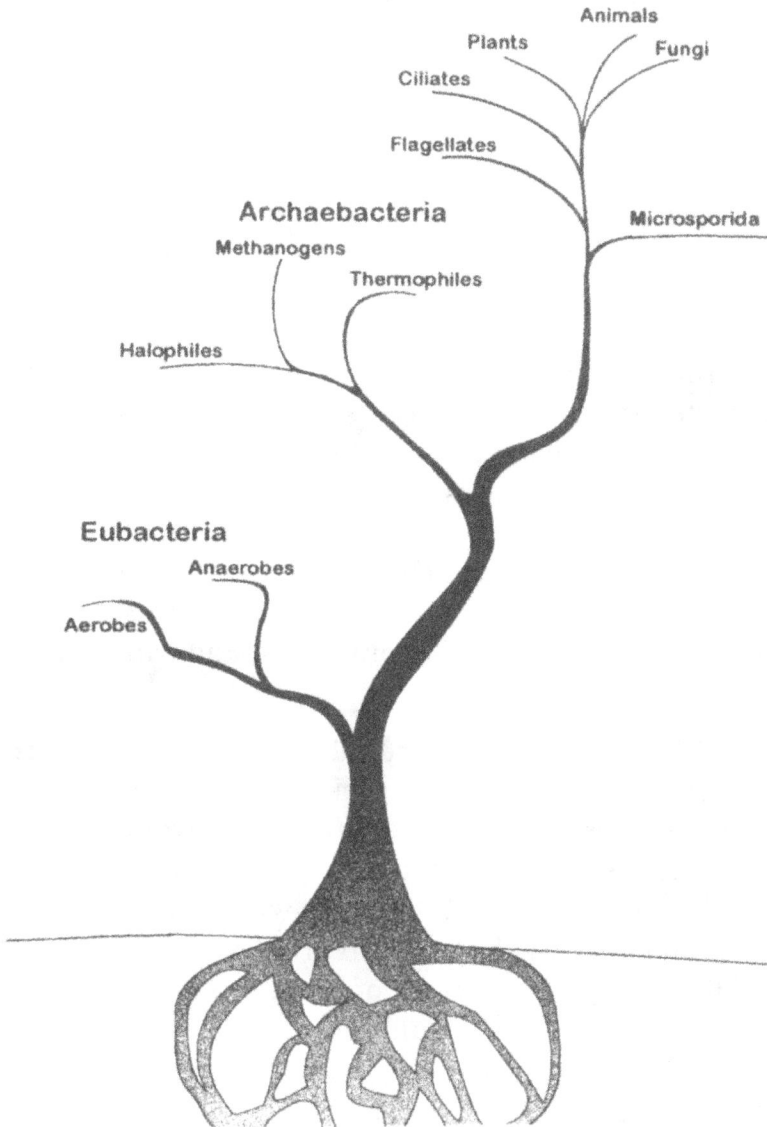

Figure 8. The tree of life.

reasonable to assume that the primogenitor itself was capable of this? We tend to use the term "primitive" for this cell, as well as its immediate descendants, but despite its original simplicity, it was, to be sure, beautiful and resourceful. All of today's cells are an extension and elaboration of that first cell.

Unless a cell can reproduce, its line will soon flicker out. Cell division is essential for its continued survival, and this means it must be able to make copies of its DNA repeatedly and accurately. Quite likely, DNA replication was at first not as refined, and editing out any mistakes less stringent. This genetic wobble might, however, have given the original cells a head start in the evolutionary race. True, large changes in the DNA could cripple the cell, but small neutral changes added to the rich variability of the gene pool, useful for meeting the contingencies of a capricious environment. This very sort of evolutionary agility must have been just what was needed in a young Earth that was itself undergoing rapid and catastrophic changes.

Without doubt, the first cell was less developed than even the simplest bacteria. Its genetic material was smaller and simpler, and more prone to copying errors. In fact, there is now good evidence that the genetic material in the earliest cell was not DNA, but RNA, which played a direct functional role in the cell, involved in the everyday activities. Bits of RNA, some of which actually served as enzymes, floating in a fluid mosaic of chemicals and encapsulated by a lipid skin, might have been the first stirrings of life.

The modern version of the tree of life with three primary branches is consonant with the concept of a simple, small, error-prone ancestor. The three branches display individual characteristics that distinguish them, one from the other. Yet, the fundamental way in which they go about their business is similar enough to indicate that they must have learned it from one teacher.

Once the first primordial cell came on the scene, equipped with its rudimentary genetic material, it was buffeted by harsh environmental conditions. Its early descendants displayed a wide range of genetic novelties, out of which three took hold and survived. All of this took place in an extremely hostile place without oxygen, perhaps in the hot sulphur springs, using the heat there to metabolise sulphur.

By looking inside living cells today, molecular evolutionists have rewritten the story of the origin of life. We may, after all, have been born in fire.

The appearance of a living cell some 3.8 billion years ago marks the most significant event in evolution. In its wake, single-celled organisms abound, triumphantly dominating the Earth for a very long time. These were cells of a simple structure with no nucleus, no paired chromosomes, no chloroplasts; they were prokaryotes. The much larger eukaryotic cells arose about a billion years ago, but they too led simple lives as single-celled creatures, such as amoeba or *Paramecium* of high-school biology labs.

Three billion years of this humdrum monotony was replaced overnight by burgeoning complexity. About 600 million years ago, multicellular organisms, the forerunners of us all, arose. They burst onto the scene with such suddenness and force that the event is referred to as the Cambrian explosion. In a matter of just a few million years, a wink of the eye in geological time, all the major groups of animals appeared. The term explosion is no exaggeration.

Much of our knowledge of this portentous moment comes from the Burgess Shale, perched high in the majestic Canadian Rockies in eastern British Columbia. The Burgess Shale contains the precious fossil records of the period just after the Cambrian explosion. The value of the Burgess Shale rests in the preservation, in exquisite detail, of the soft parts of the organisms that flourished at the time. This provides immeasurably more information that could be gleaned from the remnants of the hard parts of animals, the substance that paleontologists must contend with most of the time.

The creatures of the Burgess Shale lived on mud banks along the base of an escarpment, where they made a good living with enough oxygen, light, and heat. These very conditions should also have encouraged their rapid decay, leaving little of their soft parts. Perhaps, as the mud banks heaped up against the walls of the escarpment, they became thick and unstable, and eventually tumbled down along the slope of the mountain into the lower basins. The fauna became embedded and trapped, entombed, in the stagnant mud, where devoid of oxygen, they died quickly. In the anoxic environment, their bodies were preserved, protected from decay. Whatever the reason, the preservation of the creatures of the Burgess Shale is a godsend.

The extraordinary thing about the Cambrian explosion is the tremendous innovation that occurred in such a short period. It was not so much the quantity as it was the quality of new creatures that appeared. It was a spectacular shift from a few simple, single-celled species to many complex, multicellular ones.

Today, we recognise three kingdoms of living things. Below these in the hierarchical scheme of biological classification are phyla, which describe the major body plans of the diverse organisms. Humans and all other vertebrates are part of the phylum *Chordata*. The most populous phylum is *Arthropoda*, whose members, such as insects, spiders, and crabs, have jointed appendages. Today, there are about thirty major phyla, all of which have been present for the past 600 million years. In the Cambrian, there might have been as many as a hundred phyla, but the majority became extinct, survived by those still present today.

Two explanations have been put forward to explain the innovation of the Cambrian. The first is the ecological hypothesis. This holds that as the first Cambrian creatures came into being, there were few or no competitors. The world was overrun by bacteria, but these were probably more a good source of food than a feared seat of opposition. The newcomers filled all the ecological niches and flourished. The second is the genomic hypothesis. This proposes that the genetic packages or genomes of the species in the Cambrian were less constrained than they are in modern cells. Driven by this lax control, they underwent innumerable and major genetic changes. In the face of this propensity to genetic innovations, new forms arose right, left, and centre. In the period following the Cambrian explosion, a severe sorting process took place, and in short order many less able variants succumbed. In the millions of years since, as many as 50 million variants have come and gone.

We can trace our own origins to a small, simple, swimming creature, *Pikaia gracileus*, with which we share a notocord, the stiffening rod along the back that evolved into the spinal column of vertebrates. *Pikaia* arose in the mid-Cambrian, and from the fossil records apparently played a minor role in the unfolding story of the Burgess.

But it survived. For that, we and all vertebrates today should be eternally grateful.

Since the Cambrian, many significant events have occurred. Within each phylum, large numbers of different species emerged, but subsequently disappeared. There was tremendous experimentation, new variations on existing themes. Some made it, most didn't. The world we now know has been shaped by the unprecedented burst of innovation in the Cambrian, followed by mass extinction, and further rebuilding. For example, the Permian extinction some 250 million years ago witnessed the disappearance of almost all existing species. In the wake of this, new species emerged, but there was a striking preservation of body designs or phyla. In fact, no new phylum was formed; instead, new species arose within existing phyla to fill the gaps. More familiar is the late Cretaceous mass extinction, some 65 million years ago, in which the dinosaurs perished. In the aftermath of this, the evolution of large mammals, which so dominate the new world, became possible. The rest, as they say, is history.

Five such catastrophes visited the Earth, punctuating the history of life. The scale of these extinctions is frightening; in the Permian extinction, a staggering 96 percent of species vanished in a geological instant. In the world today, we and the estimated 20-30 million other extant species are the latest survivors.

CHAPTER 4

INHERITED WEALTH

We take a lot for granted. Surrounded by living creatures from the moment we are born, we find life so commonplace that we seldom give it a second thought. It is, however, a great wonder. The product of almost 600 million years of an upward surge in complexity, we are made of a large number of different kinds of cells — muscle cells, blood cells, brain cells, and so on. It is a dazzling feat, trillions of cells working in concert to make us who we are.

All of our many different cells ultimately come from that single cell formed by the union of a sperm and an egg. There is no obvious resemblance between this tiny cell, less than a millimetre across, and an adult. Just how does the fertilised egg find a way to give rise to the great variety of different cells, which are moulded into the various tissues and organs? How can all the complexity of an adult be fitted inside a tiny egg?

Early ideas about how this happens were not satisfying. A popular notion some time ago was that the adult was compressed, no doubt most uncomfortably, into the cramped quarters of the egg. This miniaturised person, called a homunculus, was drastically scaled down to its finest details. In this view, all that was required was simply the growth of the homunculus to achieve its full adult proportions.

Of course, this simplistic view was wrong and did not endure. It was abandoned in the mid-1800s when, with the advent of good microscopes, cells were studied more carefully, and nothing like a homunculus could be observed at any point along the line of development. What was learned instead was that growth occurred in an entirely different way. A serious inquiry into this began about a century ago, in 1894, by a German biologist, Wilhelm Roux, and his colleagues.

All higher organisms go through a similar pattern of growth initially. The fertilised egg splits into two, which in turn split into four, then eight, and so on. At first, these cells appear almost identical and form an uninteresting spherical mass. Only as the cells continue to divide and increase their number, do they begin to behave differently. Some divide more often, and their shapes change; they move about the developing embryo and take up their proper positions. All the cells eventually become equipped with specialised apparatuses, which enable them to carry out a great number of diverse functions. They become arranged in groups as tissues and organs, and work together for the good of the whole organism.

Throughout Nature, this carefully choreographed script is constantly played out. Unlike those bland single-celled organisms that have occupied the Earth for such a long time, the multicelled creatures that arose in the Cambrian lead a more colourful existence. They do not just live, grow, split themselves into two, and grow again like their single-celled cousins, but they develop. They may begin life as the single cell of a fertilised egg, but from that they develop through a complicated series of changes into a full adult.

Biologists are beginning to recognise that Nature has been economical in the tools for sculpting all of life's diverse forms. Organisms as disparate in evolutionary terms as worms and mammals use similar strategies during development. Indeed, the most fundamental of all, cell division, is the same in all living things.

Cells multiply by going through a cycle (Figure 9). During part of the cycle, the DNA replicates making a copy of the entire genome. This is the synthetic or S phase. At the end of the S phase, the cell manufactures those proteins required for its division into two; this is the G2 phase — G for gap. The G2 phase leads into mitosis or M phase, during which the cell divides into two, with each new daughter cell inheriting a full complement of the parental DNA. The end of the M phase is followed by a long interval, the G1 phase, during which new cell parts are made. The G1 phase can vary considerably from cell to cell. In some tissues, the cell moves directly from G1 into S, thus repeating the cycle. In others, it enters a resting phase, called G0, which can last for hours, days, or even years. Like the prince's kiss to Sleeping Beauty, the cell can be awakened from this dormant state and return to the cycling pattern if the need arises.

Immediately following fertilisation, the human egg undergoes a number of cell divisions, spawning a large family of cells (Figure 10). At first, an almost featureless

S   - Synthetic Phase
G_1 - Gap Phase 1
G_2 - Gap Phase 2
M   - Mitosis
G_0 - Gap Phase 0 or Resting Phase

Figure 9. The phases of the cell cycle.

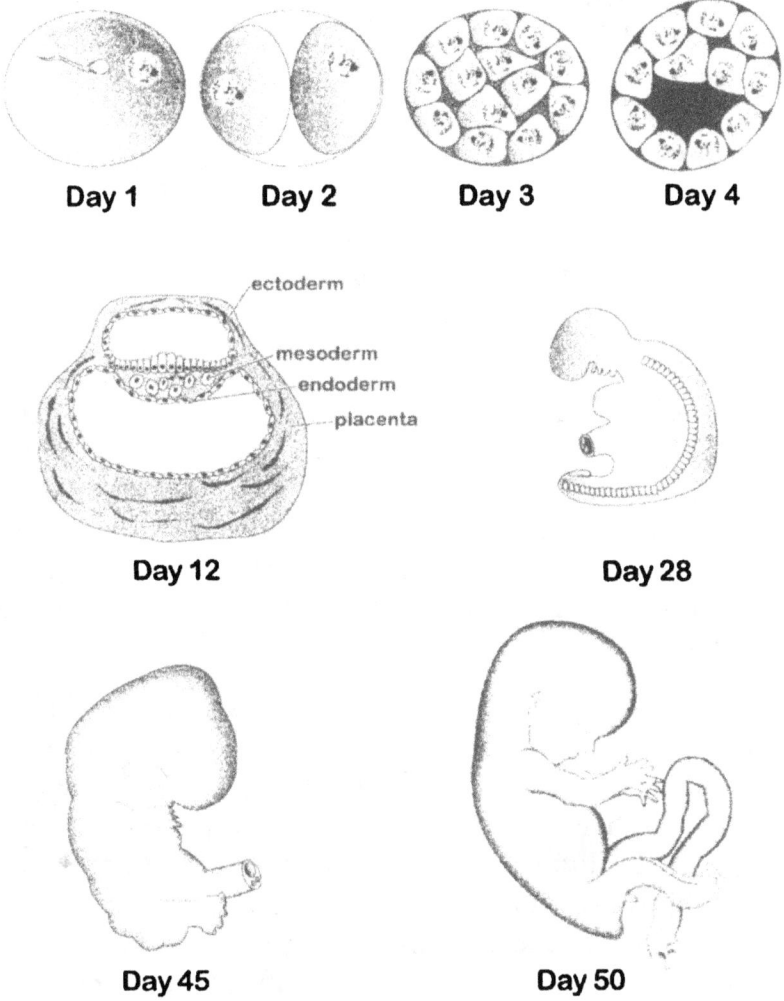

Figure 10. Stages of development of the human foetus.

cluster of cells forms, without even a superficial hint of the final outline. Soon an embryo forms. It attaches itself to the lining of the mother's womb, where it continues its development. As the embryo matures, its cells become more and more specialised and are committed to develop along certain paths. Three major stacks of cells are eventually laid down by the twelfth day. An outer layer, the ectoderm, eventually gives rise to the skin and the nervous system; a middle layer, the mesoderm, forms the muscles; and an inner layer, the endoderm, becomes the major organs, such as the intestines or lungs. In about nine months, the developing foetus reaches stage at which it can come out into the outside world as a baby. The baby grows into an infant, then a child, an adolescent, and finally an adult.

Almost all multicellular organisms have some system of development. A female frog lays her eggs on the bottom of a pond, and the male fertilises them. Soon after, tiny tadpoles are formed, swimming around and fending for themselves. Butterflies pass through a stage as caterpillars before they unfold into the beautiful creatures that they are.

Butterflies, frogs, people, they are made of large numbers of co-operating cells, each with its specific function. The taking in of food is the function of the cells of the digestive system. Terrestrial animals take in oxygen through the lungs, aquatic creatures through the gills. Everything a cell does is for the whole, a remarkable social, communal, interdependent existence. Without such co-operation, complex multicellular creatures would not have survived. It is a wonderful scheme, and one that works well. It is more powerful and productive than anything we have ever managed to build, more than supersonic jets or high-speed computers.

What underpins this complexity? We start our lives with stores of information, written as a linear sequence of bases in the DNA inherited from our parents. In a fantastic co-ordination of genetic events, bits of our DNA are read on cue at just the right moment, their instructions shared by the emerging new cells, which respond with precision and take up their position among their cohorts in the growing embryo. Nothing else in Nature comes close to the spectacle. All the instructions for what we are and what we become are contained in the DNA of the fertilised egg.

Multicellular organisms are vastly more elaborate than single-celled ones. They require much more genetic information for living, and it should, therefore, come as no surprise that humans have 46 chromosomes compared with only one in *E. coli*. Moreover, the human chromosomes are larger, storing incredibly more bits of information.

Higher organisms have several more kinds of cells than simpler ones. For instance, bacteria come in two cell types; yeasts three; hydra 13 to 15; some worms 60; and humans 250. As we climb the ladder from simple to complex creatures, we find that the quantity of genes increases. This is exactly what we would expect. But when we look at the amount of genetic information as we go up that ladder of complexity, we come across an interesting pattern. Stuart Kauffman found that as the number of different cells in an organism increases, the amount of DNA also increases, but it does so at a faster rate; for each new cell type, the DNA doubles (Figure 11). This means that in the course of evolution as organisms became more complex with more and more different kinds of cells, acquiring a new cell type requires ever greater amounts of DNA.

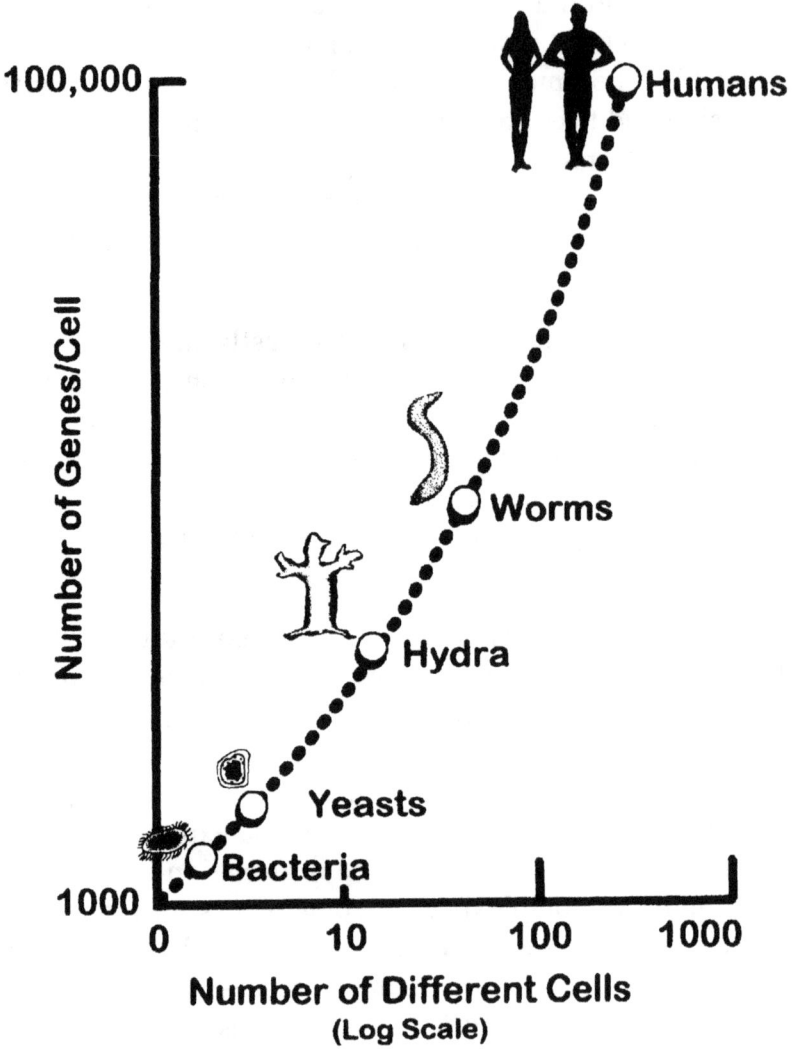

Figure 11. Graph of the relationship between the number of cell types and the amount of DNA in different organisms.

It is germane to ask why progressively more genes are required for each new additional cell type. True, humans have a hundred-fold more cell types than bacteria or yeasts, but the problem is more than just the number of cells. In humans, cells are organised into elaborate organs, like the brain or eye or pancreas. In each of these, there are distinct cells, which come on stream at the right moment as the developing embryo matures, taking up their station and beginning to function. The intricate series of steps that must fall precisely into place in any higher organism if it is to be born at all, is all ordained by its genes. It is this co-ordination of meshed events as the development of the embryo unfolds that requires the extra DNA. The embryo follows a genetic blueprint, which ensures that every tissue organises itself properly and not just any old way. This is the marvel of development.

After birth, the exquisite co-ordination of events continues unabated. Our fingers grow to about six centimetres, our legs develop to carry us, and our thorax changes in proportion to the adult dimensions, and so on. When the time comes for growth to stop, it does. There is no room for ambiguity, no wandering from the rules agreed to at conception. The journey that begins in the embryo goes on through life. All of this is plotted on the genetic map of DNA with its thousands of genes, its millions of base pairs.

The sheer number of cells that take part in this truly marvellous plan is mind-boggling. An adult is made of fifty trillion cells, all arising from the single fertilised egg. As each cell gives rise to only two cells at a time during cell division, it takes a tremendous number of generations of cells to achieve this. Moreover, it is from this original cell that all of the 250 different kinds of cells come.

At first, the early embryonic cells can develop into any one of the many cell types, but as the embryo grows, sooner or later, the broad potential of any individual cell becomes restricted to a specific cell lineage. The blood cells form a lineage with a variety of cells, each catering to a particular function. The red blood cells transport life-sustaining oxygen throughout the body; the tiny platelets aggregate at the site of a cut and seal it off to prevent bleeding; the white blood cells make up the immune system that defends us against any microbial intrusion. All of these cells develop from a special cell, the stem cell, which, in the case of the blood cells, resides in the bone marrow.

All organs have a small pool of stem cells, maybe less than one percent of the total cell number, that can divide over and over. When a stem cell divides, one of the daughter cells remains as a stem cell, while the other divides further into a specialised kind of cell. This process of becoming specialised is called cell differentiation. As the cells differentiate, they become committed to a particular lineage, and eventually can form only one kind of cell.

The stem cells come from the original fertilised cell, and are laid down early in development. All cells arise from this stock; hence the name stem cells. Once established, the stem cells recreate themselves through their self-renewal capacity, in which one of the daughter cells also becomes a stem cell, thus maintaining the pool. The other daughter cell differentiates into a specialised cell; it undergoes about 10 to 12 cell divisions, and after a limited life span, it dies.

There is a constant, and in some tissues an astonishing, rate of cell turnover. As old cells bear the stress of wear and tear, they die, and, as they do, new cells move in to take their place, in the same volume, cell for cell. On average, a person creates 3 to 10 billion platelets, red cells, and white blood cells every hour. This goes on all the time. When there is an increased demand, such as the result of a bleed, cells are produced even more rapidly, sometimes up to ten times as fast, to deal with the emergency.

Under normal conditions, the red blood cells live for about 120 days, after which they become trapped in the spleen, where they are torn to bits. As this happens, new red blood cells leave the bone marrow to take their place. The platelets stay in the blood for a week, and then they too break up, making room for younger platelets. Some white blood cells, the neutrophils, travel around in the blood for just a few hours; others, the lymphocytes, live for a longer time, years or even a lifetime.

An organism contains many kinds of stem cells. Some generate the various blood cells, some the cells of the skin, and so forth. In a sense, the stem cells are immortal during the lifetime of an individual. The specialised or differentiated cells, however, live and let live. As they age, they lose their zest, and at prescribed times, they die, replaced by younger cells; in a way, we are continually being renewed.

As the fertilised egg develops, it spawns a growing family of cells. Quite early, the cells have an unlimited repertoire, each capable of generating a complete adult. As development proceeds, however, and the three layers are laid down, this omnipotent potential is lost, and cells become restricted to one or another lineage. The ectoderm, for instance, gives rise to the cells of the skin and nervous system. Later, the range becomes even more restricted to specific kinds of cells.

In animals, development occurs in a succession of stages. Once the layers of cells are set down, an important event occurs: the main axis becomes defined in the embryo. All animals — worms, butterflies, and mammals — have the same body axes — head and tail, front and back, and left and right. It is an obvious point, but a real challenge to the nearly spherical embryo. Evidence is at hand that embryos from quite diverse species, some that have diverged at least 600 million years ago, use a single molecular map to plot their co-ordinates.

The axis of the embryo is laid down by the manner in which some maternal messenger RNAs in the egg are translated. The egg, larger than the sperm, contains a store of messenger RNAs from the mother, which it stashes away, silently, until fertilisation. Although these maternal messenger RNAs are distributed evenly, they are translated into proteins only in the cells of the "head" region, so that a high concentration accumulates there, whereas in the "tail" region, where translation is blocked, little of the protein is present. The variation in the protein levels between cells in different parts of the embryo is, in turn, responsible for activating different genes. Cells which to this point have been identical, now are different.

This idea of a gradient in the concentration of proteins across the developing embryo was first proposed by Lewis Wolpert. In essence, certain key chemicals can tell the cells where they are, and by looking up their DNA manual, find out what to do. In this way, the single fertilised egg develops into the different kinds of cells.

After the initial prompting by the maternal RNAs, once the ground plan is sufficiently laid down, further development of the embryo comes under the direction of its own DNA. Proteins appear and disappear as different genes are switched on or off. This demands an extraordinary co-ordination of a number of genes during development. Cells are given instructions when to divide and what to become, sculpting the tissues, and giving them their final form. In the end, the simple, one-dimensional array of bases in DNA is transformed into the complex, three-dimensional organisation of the embryo and, later, the adult.

During the period of embryonic growth, the total cell number expands enormously as the fertilised egg first divides into two, which in turn divide into 4, then into 8, 16, 32, 64, 128, 256, 512, 1024, and so on, in a geometric progression. After ten cycles of cell division, the single cell spawns about a thousand descendant cells. Cells move from one cycle to the next without pause. Growth occurs at an accelerated rate, faster than at any other time in life. In addition to this explosion in cell number, the degree of complexity of the cells also increases substantially. The cell's machinery is operating at full throttle during this period, and the demands on the DNA are tremendous.

How does the embryo keep the process rolling? To frame the question another way, how does the genetic material rise to the challenge in this demanding period of growth and development? There is not a single answer to this, but it appears that different organisms face the problem in more than one way.

A good deal is known about how the DNA of a frog's egg becomes cranked up to full speed. In the course of the first few weeks of growth, a trillion ribosomes are needed, and this requires the production of vast quantities of ribosomal RNAs and ribosomal proteins. The number of genes for the ribosomal RNAs increases a hundred- to a thousandfold to supply the material for the massive stock of ribosomes. This does not come from an excessive multiplication of entire chromosomes, but the relevant genes are snipped from the chromosomes and their numbers amplified. Groups of genes from one to 20, arranged in tandem, are extruded from the main chromosomes and form small circular DNA molecules within the nucleus. From these, ribosomal RNAs are rapidly transcribed in copious amounts.

The production of ribosomal RNA is only one step. A correspondingly high rate of transcription for the messenger RNAs coding for the ribosomal proteins must also occur. To achieve this, the DNA, which is usually compact, becomes extended into long loops. The chromosomes take on a bizarre shape, called lampbrush chromosomes. The name comes from the resemblance to the old lampbrushes which were used to clean soot from the globes of oil and gas lamps. In the normal compact state, DNA is secluded, but when drawn out into loops, it is exposed, allowing free access to its information, which can be copied at nearly maximal speed. New messenger RNAs emanate at a rapid rate, unmatched in the ordinary course of events.

During development of the amphibian egg, another mechanism ensures that proteins are synthesised quickly and in large quantities. Single genes are not enough. Instead, the required genes are duplicated, or amplified, about six times, giving rise to 64 copies ($2^6 = 64$). Unlike the snipping of certain genes from the chromosomes and their subsequent multiplication, the duplicated genes in this situation remain an integral part of the chromosome, giving it an "onion-skin" appearance (Figure 12).

The picture emerges that the genetic apparatus is flexible, adaptable. Yet, in a truly amazing feat of

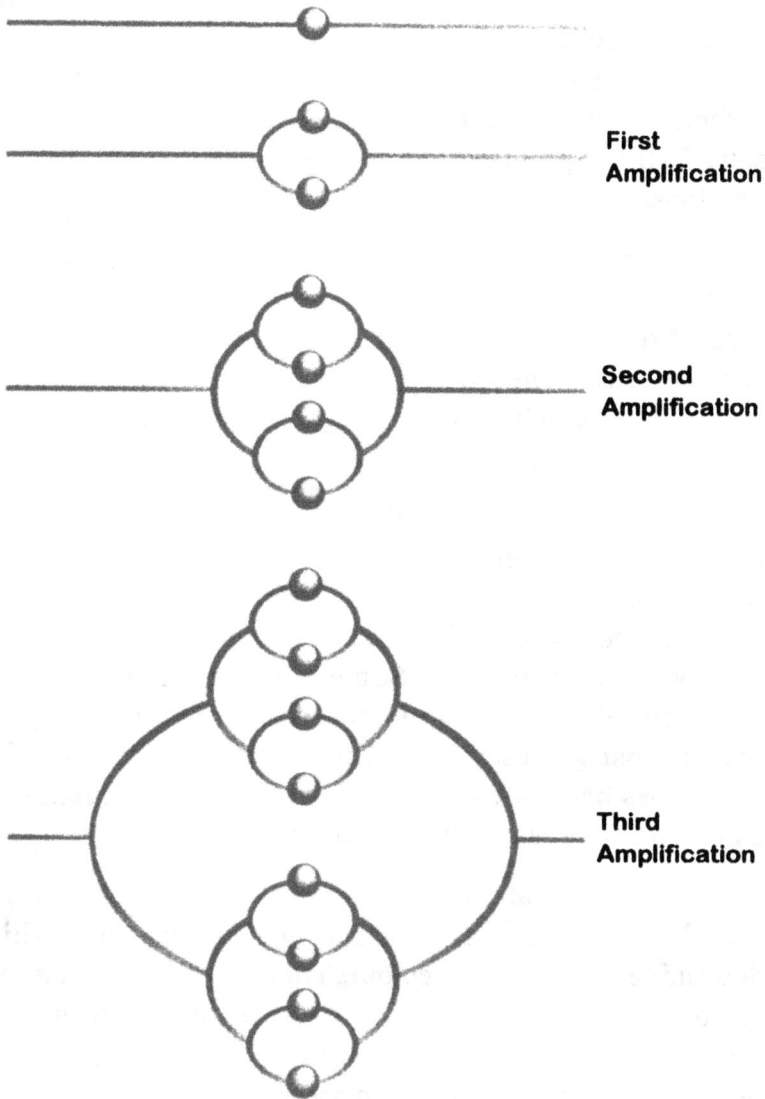

**First
Amplification**

**Second
Amplification**

**Third
Amplification**

Figure 12. "Onion-skin" appearance of the chromosome
due to amplification of a gene.

development, the cells manage to arrange themselves into a complete organism. An insight into how this is achieved came through the efforts of geneticists, who built up a huge library of genetic aberrations in fruit flies between the 1920s and 1980s. It is often the case in biology that bizarre defects in some odd organism provide valuable clues to solve baffling problems. So it is with the way the fruit fly develops.

Some of the genetic defects in fruit flies are relatively minor, such as the colour of their eyes or the type of their body hairs. Others, however, are much more serious, and they disrupt the whole developmental programme, which is often brought to a grinding halt.

Edward Lewis at the California Institute of Technology studied a small cluster of genes in fruit flies, called homeotic genes. The name comes from the Greek word "homeo" meaning "alike." A mutation in one of the homeotic genes caused one body segment to change into the likeness of another. Some variants of the fly were found in which their antennae were replaced by legs; it was a most grotesque mutation, termed antennapedia. Other flies had a second pair of wings due to a mutation in a homeotic complex called bithorax.

The special feature of the homeotic genes is that they behave in a more abstract way than genes which determine traits like eye colour. They tell the cells that they are part of the fly's head, thorax, or abdomen. A mutated antennapedia gene complex can produce a perfectly normal leg, but in the wrong place.

Homeotic genes change how we look at development. It now seems that the process involves variations on a theme. For example, the homeotic gene complex antennapedia assigns the developmental plan to the thorax during all stages of growth — embryonic and pupal — even though different structures, such as sensory organs, legs, or wings, develop in succession along the thorax.

Homeotic genes program not individual pieces, but how those pieces are organised. They serve as crucial switch points in development. The proteins of these genes bind to DNA at specific regions, where they turn on or turn off large groups of subordinate genes. The crux of the matter is they regulate other genes.

We now begin to see the outlines of a regulatory system in the embryo. So far, we have been concerned with the genes that code for proteins which provide the scaffolding for cellular structure or which conduct cellular metabolism. Behind the scene, however, are important genes that regulate when these are called into service and for how long. The architecture of the genomic regulatory system is slowly being built up. The group of homeotic genes forms a strong pillar of that system.

In the past decade, genes strikingly similar to the homeotic genes of fruit flies have been found in the chromosomes of many animal species, frogs, mice, and people. The obvious inference from this is that these genes share a common evolutionary ancestry that goes back millions of years.

Although the structural similarity of the genes in a fruit fly and a mouse argues for a common origin, it does not necessarily mean they serve the same developmental function in their respective embryos. The span of time since these species diverged and followed different evolutionary paths could give the genes ample opportunity to evolve new abilities. However, it is now becoming apparent that the protein of the mouse gene can substitute for the mutant protein in the fly. This means that these proteins convey a similar message to early embryos regardless of their final form. It seems that the system for guiding the embryo through its early development came on the scene a long time ago. This ancient system has proved so useful that animals still rely on it today to map their body design.

CHAPTER 5

GUARDIANS OF THE GENOME

We share the planet with countless other species. They have, as far as we can remember, always been with us — friends and foes. In the modern world, we have gained the upper hand and, for the most part, are oblivious to those creatures in the open. We think of them as a kind of abstraction, largely hidden from our everyday view. We have come to be comfortable with a few that we keep as household pets, and occasionally some invoke an upwelling of such emotion that we march in the streets on behalf of their welfare.

But germs are another matter, quite another matter, indeed. They surround us, invisible, but hard to ignore. We are constantly reminded of their presence in advertisements all around us, but especially on television.

We spray disinfectants everywhere, pour them down our kitchen sinks and toilets, and scrub our floors dutifully to keep these microbial invaders at bay. We are cautioned to be vigilant in our ceaseless fight against the teeming throngs which seem forever after us, missing no opportunity to invade our bodies. Sometimes, we seem so caught up in a paranoia that we forget that most microbes are, by and large, harmless.

In real life, the vast majority of the microbial world is totally uninterested in infecting us. Even those browsing on our skin or living in our intestines are rarely harmful. Infection is the exception. To be sure, though, there are some that are a real threat, to be avoided at all costs. Given the chance, they invade and gain access to our deepest tissues, where they continue their destruction, eventually spilling over into the blood and causing the most violent reactions.

These microscopic adversaries have always been with us. Occasionally, they strike with such ferocity that we recoil in fear. The unexpected surfacing of the Streptococcal "flesh-eating" bacteria in our cities or the sweeping rampage of Ebola virus in deep Africa reminds us of the potential threat. For a time, we become fearful, worried that we are on the verge of being overrun by a modern microbial demon.

The confrontation between the microbial invaders and us is as old as life itself. During the long course of evolution, we have become equipped with elaborate systems of defence, our own internal Pentagons. Of these, the immune system is the most complex, indeed the most bewildering genetic mechanism in biology. Many types of cells take part in this defence, and many genes are devoted to it.

The immune system has been recognised for a long time. It is well known that after we recover from several bacterial and viral infections, even going as far back as childhood, we become immune or resistant. We do not come down with the illness again, even if we are in close contact with an infectious person at some later time. Children who contracted chicken pox will not suffer from the disease again, even as adults, when they become exposed to this virus. And so it is with many other childhood infections.

This is the result of the marvellous property of the immune system to protect us. It has the ability to remember for long periods, sometimes a lifetime, previous exposure to many germs. When faced with the microbe again, it reacts quickly, turning on every arsenal at its disposal; it seals off, blockades, and neutralises the offending intruder before it can do any harm.

A major part of our immune system is the antibodies, special proteins made by lymphocytes, a subset of our white blood cells. Antibodies stick to certain molecules, collectively called antigens, on the surface of microbes. For each antigen there is a matching antibody that will latch on to it. A mysterious feature of the immune system is its capacity to produce an appropriate antibody whenever it is exposed to a new antigen. When we come across a germ with its own peculiar set of antigens on its surface, we make specific antibodies to those antigens. The memory of this encounter is stored, and when a second challenge arises, we react with speed and precision to stave off the attack.

We are born with the potential to make a vast set of different antibodies. Which ones we end up making depend on which germs we come in contact with during our lifetime. The repertoire of antibodies is truly wide-ranging. For practical purposes, we can make 10^6 to 10^8 different antibodies. This is fiendishly large number that would require an enormous number of genes devoted for just that purpose.

Even more fascinating is that antibodies can be made not only against antigens present on other living things, but against manmade molecules that never occur naturally until chemists synthesise them in their laboratories. During the long course of evolution as organisms developed, they devised this remarkable strategy to protect themselves against predators and intruders. The ability to make specific molecular missiles, antibodies, that are launched against molecular targets, antigens, on the surface of the offenders is itself a clever accomplishment. But it seems to go beyond that. The system is versatile enough to respond to artificial substances as well.

The general structure of antibodies is simple enough. They are built according to the same master plan, but contain sections that differ from one another. Each antibody is made of two identical pairs of proteins — two small light chains and two large heavy chains. The four chains are assembled in the form of the letter "Y" (Figure 13). They are held together in the foot by chemical bonds between the heavy chains, whereas the arms are formed by a hinge in the heavy chain. The light chains combine with the arms of the heavy chains.

The sequences of amino acids in the tips of the arms of the light and heavy chains are different for each kind of antibody. This gives each antibody its special unique identity. The top part of the molecule with its different

Antibody Molecule

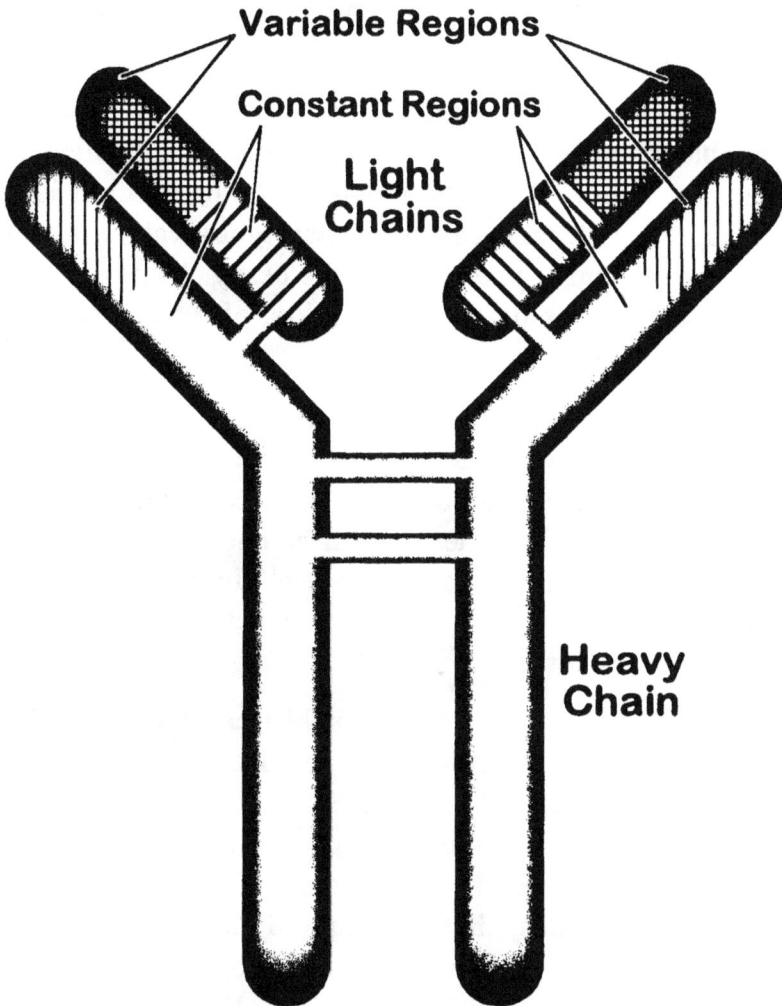

Figure 13. Y-shaped structure of an antibody molecule.

amino acids is dubbed the variable region, and it creates the surface contour necessary to dock with an antigen. Whether an antibody sticks to an antigen depends on the "fit" between the two molecules. If there is a close match between them, they come together; otherwise, they drift apart, and the antibody sniffs out other available antigens for a proper match.

The rest of the antibody molecule below the variable region is more or less the same, and is called the constant region. The lower half of the light chain is its constant region, and can be of two types, called lambda and kappa. Similarly, the part of the heavy chain below the variable region is its constant region, of which there are nine types.

The generation of the millions of different antibodies, each with its unique variable region, presents a problem. Are there millions of genes, each coding for an antibody? If so, then there would be hardly any room left in the chromosomes for anything else. This is just the start. Each antibody requires at least two genes, one for the light chain and one for the heavy chain.

In the conventional thinking about how DNA works, the explanation for the wide diversity of antibodies is that a separate gene codes for each. The Y-shaped structure of the antibody with the arms made of two separate proteins allows some relief. Let us assume one million antibodies; then, we should expect at least one million genes. However, if we suppose that the variable region of the light chain and that of the heavy chain contribute equally to the unique configuration of each antibody, then the problem becomes less formidable, although still biologically implausible. By assuming an equal contribution from the two variable regions, then a thousand genes each for the light chains and heavy chains would yield one million different antibodies — 1,000 light chains can combine with 1,000 heavy chains in a million possible ways: 1,000 x 1,000 = 1,000,000. This is the minimum number of genes. Much less than a million, but still a great deal for a cell to carry.

In the 1960s, biologists were quite aware of the dilemma in figuring out how the immune system made this large number of antibodies. The traditional biology of the time could not explain it, and, indeed, the very existence of the immune system seemed a contradiction. Not surprising, therefore, as soon as the techniques of molecular biology became refined, the genes of antibodies were among the first to be studied. A pioneer in the field is Susumu Tonegawa who, with his colleagues at the Basel Institute of Immunology in Switzerland, made a significant contribution to our understanding of these genes. For his discovery, Tonegawa was awarded a Nobel Prize.

The new biotechnology made it possible to count the number of genes in a cell. When this was done for the genes for antibodies, the first surprise was that much fewer genes are devoted to their production. There are only about 300 genes for the variable regions of the light chain, and a similar number for those of the heavy chain. These are considerably less than predicted. Another unexpected discovery was that the genes are in pieces. The segments of DNA coding for the constant and variable parts are not together, but are separated in the chromosome by a small stretch of DNA, which codes for no recognisable protein.

What this means is that there is no gene, in the conventional sense, for the protein that makes up a complete chain. Instead, there are separate segments of DNA for different portions of that protein. Yet, in the antibody molecule, the variable and constant parts form a single protein chain.

We had come to think of a gene as a continuous stretch of DNA that codes for a complete protein. Now, however, this notion is shot by what we have learned about how antibodies are made. Two separate DNA segments, each coding for a part of the protein, are involved, with the pieces coming together and joining up to form the whole. Co-ordinating this is not easy. The two pieces must be made in equal amounts, at the same time, and then brought together.

This discovery of the unusual nature of the antibody genes was the first description of a "patchwork" pattern of genes. Not long after, it was realised that the genes of higher organisms are typically organised in this manner; rather than a continuous, uninterrupted stretch of DNA, there are intervening segments which do not code for amino acids in the final protein molecule. It is not that this overturns everything we have known about genes, but it does point out how things that used to seem clear and certain must be amended, revised, modified as we gather more information. The mystery behind the generation of antibodies has revolutionised some of our concepts of genes.

The news of the patchwork pattern of genes was followed by an even more remarkable discovery. In embryonic cells, there are three separate DNA segments on the same chromosome for the light chain: a segment V for the variable region, a joining segment J, and a segment C for the constant region. Each light chain, then, is the combination of V-J-C. The V segment is separated from the J segment by a considerable space, and between the J segment and the C segment there is also a smaller space. Cells with this DNA structure are unable to make a full-fledge antibody with the result that the foetus and even the new-born cannot produce their own antibodies. At birth, the baby is protected for some time by antibodies it gets from its mother.

Soon after birth, the mechanisms to make antibodies are put in place. This is shown in Figure 14. In each lymphocyte, the first step is taken toward production of light chains by excising the piece of DNA between the V and J segments, and so bring these two together. The C segment remains separated at some distance from the new V-J complex. This entire portion of the DNA is transcribed into a messenger RNA, and only at this stage is the piece between the V-J and C segments cut off. In the end, the messenger RNA contains a copy of the V-J complex and the C segment. These form a single continuous RNA strand, which is then translated into protein by the ribosome to make a complete light chain.

There are two gene systems on separate chromosomes for light chains. One is the lambda gene system, which contains about 300 V segments, six J segments, and six C segments. The other is the kappa gene system, which, likewise, contains 300 V segments, five J segments, and a single C segment. A lymphocyte generates either a lambda or a kappa light chain, but not both.

The assembly of the genes for the heavy chains occurs on another chromosome. It follows a similar pattern, but in addition to the three segments (V-J-C), there is a fourth segment called D for diversity. The D segment

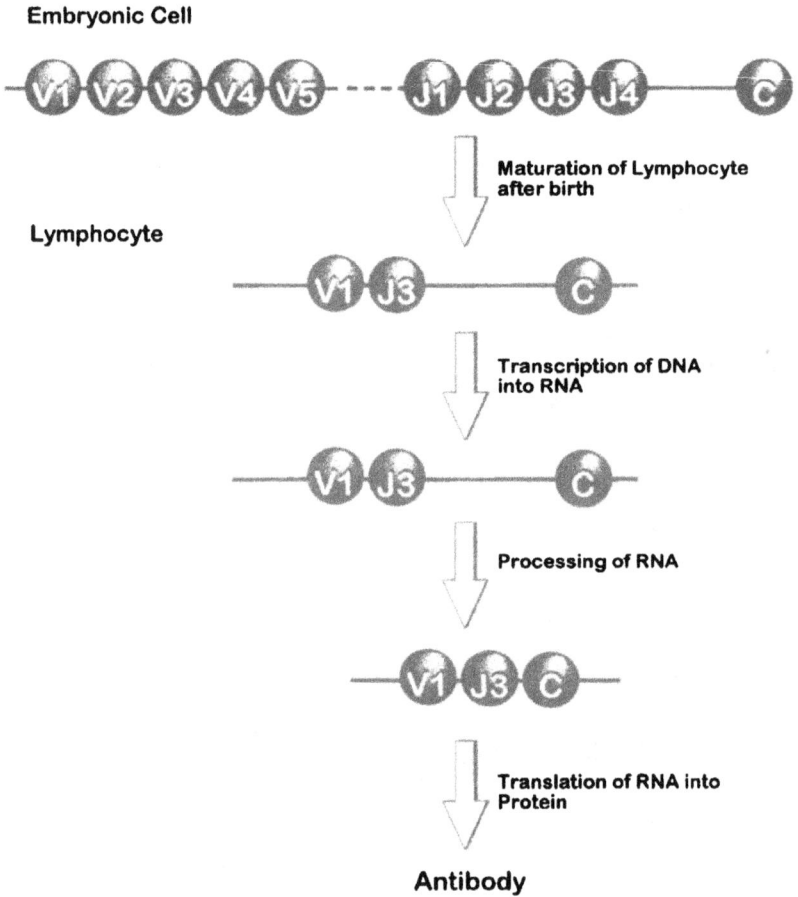

Figure 14. Genetic mechanisms of antibody production.

was discovered when an extra two to thirteen amino acids were found in the protein between the V and J segments. Although relatively short, the D segment plays a crucial role in the diversity of the antibody.

The genes for the heavy chains are assembled by the sequential joining of the different segments. First, the D segment is joined to the J segment; following this, the V segment is joined to the D segment. The gene system for heavy chains is contained in a single cluster. There are about 300 V segments, next to which are about 20 D segments, followed by four J segments. Downstream lie all the nine C segments.

With all these different genes, what can we predict of the diversity of antibodies? During maturation of the lymphocyte, some interesting things occur. It appears that all, or almost all, V and J combinations are tried out in the process. This creates the opportunity for a large variety of different antibodies. For instance, by combining any one of the 300 V segments with any one of the six J segments, some 1,800 different light chains can be constructed. Likewise, for the heavy chain, the 300 V segments, 20 D segments and four J segments, can yield almost 24,000 different combinations. As antibody molecules are formed by the assembly of the light and the heavy chains, an enormous number of different antibodies, in the millions, can be put together.

This generation of diversity from the assembly of different light and heavy chains is impressive enough, but there is more. When the base sequences of the antibody genes in embryonic cells are compared with those in lymphocytes, some curious discrepancies are observed. It appears that at some point in the maturation of the lymphocytes, additional mechanisms introduce further genetic changes. These, as it turns out, are brought about by mutations in the antibody genes, programmed errors as a means to augment the variety of antibodies.

That mutations are used by Nature to expand the variability of antibody molecules called for a shift in our concept of the way DNA works. Far from being obstinately inflexible, sticking to the point, DNA is turning out to be surprisingly fluid, sometimes chancing it all. The genome is proving to be unimaginably versatile, adaptable, and constantly adapting. It is a hard idea to accept. We prefer to imagine that over the aeons, the genetic apparatus has gone to great lengths to insure itself against changes; yet, it now seems that the DNA molecule is ordained to undergo changes.

Not long ago, we believed that the DNA blueprint for an organism was fixed as soon as the egg was fertilised. It would remain that way, unchanged, throughout life. Cells read their own different parts of this programme and followed the instructions. The whole scheme was a rigid, ordered one. Of course, we accepted that mutations occurred from time to time, but looked at this as a mere accident, something dropped, to be avoided at all costs. We see before us now Nature's ingenuity in deliberating using mutations to make millions of antibodies from a meagre set of original genes. It is another of those astonishing discoveries of modern biology that leave us in awe.

Most of the mutations occur at the V-J or V-D-J junctions of the antibody genes. A few bases are inserted or removed, altering the sequence of bases, which, in turn, dictates the sequence of amino acids. In addition to changes at these junctions, other mutational events can occur upstream in the variable region.

What of the lymphocytes that make these antibodies? They roam through the tissues and blood, continuously monitoring. When a germ strays into the body, its antigen is sensed. A lymphocyte with specific receptors on its surface hunts down and engages the invader. Immediately after this encounter, the lymphocyte changes. It enlarges, makes new DNA, and undergoes a rapid series of divisions. A large family of cells, each with the same receptor that can connect to that foreign antigen, is spawned.

How could we anticipate which germ will invade? We do not, of course, know ahead of time, but have the mechanism in place to respond to any such intrusion by making antibodies to an enormous variety of germs. The lymphocytes seem informed about everything foreign around them. Because there are so many of them, they can make collective guesses at almost any alien microbe that wanders into unauthorised territory. The lymphocyte with the proper receptor, which binds to the microbial antigen, is then picked from the millions of others to take care of the threat.

The plan of defence is already in place, the weapons chosen and primed to fire. Once the word is sent out, a fierce showdown ensues. The white cells are called in from everywhere, and they seal off, entrap, and finally engulf their prey, which is ripped apart into tiny pieces by potent enzymes. When it is all over, the lymphocyte resumes its patrol. It retains the memory of the confrontation, and responds to a similar threat at any later time with stunning speed and efficiency.

Meanwhile, the other lymphocytes remain quiescent, biding their time until they are called into service. And then, they too do their work, compulsively and automatically.

CHAPTER 6

GENES OF WRATH

C opernicus wrote his observations of the movements of the heavenly bodies in a tract entitled "De Revolutionibus Orbium Caelestium." The word revolution in this original sense denoted the circular movement of celestial bodies. Over the centuries, it has come to mean the overturning of conventional order. The discoveries in the past decade about the origin of cancer have been nothing short of revolutionary, for they have put our concept of the disease into a different orbit.

There is something unnerving about all this, since it is now evident that cancer is caused by the malfunction of genes. Further, it turns out that these are key genes, which also play a vital role in normal cell regulation. Without them, the organisation of higher organisms would not be possible; yet, they are sometimes responsible for the wracking and unhinging of everything.

The human cell contains many genes that work in unison to co-ordinate normal development, deciding when a cell should split into two, and what type of cell it becomes. When these genes are damaged, the fine regulation is in shambles; order is lost, and out of this chaos a cancer arises.

We have, it seems, always had to deal with cancer. Its dreaded signature has been inscribed in some of our oldest written records. The Ramayana, the Ancient Indian Epic of 2000 BC, alludes to a disease much like cancer, and the Egyptian papyruses of 1500 BC contain records of malignant tumours. The problem was also recognised by the ancient Greek physicians, Hippocrates and Galen, who, in the tradition of their time, believed that harmony between the "Four Humours" — blood, phlegm, yellow bile, and black bile — was essential for health. According to this view, a cancer developed when there was an excess of black bile or melanchole. Cancer was, indeed, a melancholy disease.

Over time, as theories came and went, doctors described the clinical aspects of this terrible disease more fully, but could offer no reliable explanation about what caused it. The first step towards such an understanding came in the mid-1800s with one of the great advances in biology: the enunciation of the cell theory by Mathias Schleiden and Theodor Schwann. In this, they described the universal presence of the cell in all plants and animals.

The cell theory was brought to its triumphant maturity by Rudolph Virchow, a German pathologist. He wrote extensively on several topics on health and society, but his lasting contributing was his diligent and careful description of the central role of the cell in disease and in health.

Another German, Theodor Boveri, ushered in the new era of cancer research at the turn of this century. Boveri took a great interest in chromosomes, and realised that they were indispensable for the normal growth and development of an organism. He recognised that the full set of chromosomes was necessary for growth to take place properly; an unbalanced set caused abnormal growth. He took this idea further to suggest that a cancer resulted from an abnormal chromosome set. A normal set was necessary for the normal functioning of a cell; if a single chromosome was absent, the cell would be defective, and it died. He speculated that there might be chromosomes which inhibited cell division; loss of these would lead to unbridled growth. Alternatively, there might be chromosomes which promoted cell division; an excess of these would cause rapid cell proliferation. To Boveri, a delicate chromosomal balance was essential; any disturbance of this would cause a cancer.

The essence of Boveri's concept became generally accepted, but final proof came from the most unlikely source — viruses. In 1908, Vilhelm Ellerman and Oluf Bang speculated that a virus might cause leukaemia in chicken. Acceptance of this was obscured by two facts. First, the idea that a virus could cause a cancer was frowned upon, and, second, leukaemia was not even considered a cancer at the time. Two years later, Francis Peyton Rous described a cancer, called sarcoma, also in chicken, which he thought was caused by a virus, later named the Rous sarcoma virus.

Considerable scepticism surrounded much of the early work of the viral origin of cancer, but slowly the idea became accepted. In 1966, Rous was awarded the Nobel Prize for Medicine and Physiology for his part in the discovery of the tumour viruses. There was a lapse of over fifty years between his discovery and the award of the Prize, the longest in the history of the Nobel committee. It took a long time for the idea to enter the mainstream of biological thinking.

The study of the RNA viruses went a long way in unravelling the secrets of cancer. Some of these, the retroviruses, infect cells in which their genomic RNA is released. From this, a double-stranded DNA molecule, the provirus, is made under the direction of a special viral enzyme, reverse transcriptase. The provirus then elbows its way into the cell's chromosomal DNA, where it looks just like any other piece of DNA. The proviral DNA is transcribed like any other gene, and its message, coded in its special sequence of bases, is translated in the cell.

The virulence of the retroviruses was a valuable clue in understanding how they caused cancer. Some viruses induced cancer in animals very quickly, while others seemed less efficient and did so only after relatively long periods, usually several months. In 1970, Peter Duesberg and Peter Vogt found that the RNA genome of the Rous sarcoma virus was longer than that of an ordinary retrovirus. This immediately led to the speculation that the extra piece of genetic material was responsible for the efficiency with which the Rous sarcoma virus inflicted harm. This is exactly what happens. The Rous sarcoma virus contains an extra gene, dubbed the *src* gene for sarcoma, which can derail normal cellular function.

Not long after this important lead was provided, a vigorous search for the putative cancer gene began. Harold Varmus and Michael Bishop at the University of California in San Francisco looked for the *src* gene in healthy chicken. The result was startling. The chromosomes of normal chicken cells contained a *src* gene. The very gene that was responsible for a cancer when it was inadvertently dragged into a normal cell by a retrovirus, could be found in that cell. Not only was the *src* gene found in chicken cells, but a similar gene turned up in other animals, as well as in humans. It seemed that wherever one cared to search, the *src* gene showed up.

It now appears that the retrovirus picked up its *src* gene from animal cells at an earlier time in evolution. Once this gene has been grafted into the viral genome, it is tugged along from one animal to another, sometimes with grievous consequences.

We have to face, in whatever discomfort, the real possibility that all normal cells have genes with the potential for causing cancer. We carry pieces of DNA in our cells, some of which resemble the viral cancer-causing genes, and may well have been copied and grafted into the viral genome from us at some remote time. The discovery of potential cancer genes, called proto-oncogenes, in our cells was a seminal event, and in recognition of their pioneering work, Bishop and Varmus were awarded the 1989 Nobel Prize for Medicine and Physiology.

In the 1980s, the proto-oncogenes led to a radical change in thinking about the origin of cancer. The story is being pieced together. The retroviral oncogene, such as the *src* gene, that induces cancer in animals originates from the proto-oncogene of animals and humans. Once it is taken up by the virus, it becomes modified in a small but critical way, yet not so much that the family resemblance is lost. When it re-enters a cell, the viral oncogene has a devastating effect: there is rampant cell division, setting up the conditions for malignant growth.

But what can we make of the cellular proto-oncogenes? What do they do in our cells? Since the description of the *src* oncogene, about fifty other proto-oncogenes have been identified. When we consider that the typical human cell contains over 100,000 genes, this is a relatively small number.

The clue to their possible function comes from the discovery that distantly related species have similar proto-oncogenes. The ancestors of these genes have been passed on, conserved, through the entire course of evolution, undergoing slight changes over the vastness of time, but their fingerprints leave no doubt of their identity. The very conservation of the proto-oncogenes across these boundaries attests to their vital role in the life of cells. They are, in reality, the regulatory elements for the growth and development of living organisms.

Just how important the proto-oncogenes are can be easily gleaned when we consider the remarkable chain of events that goes into the making of an individual. A single fertilised egg rolls its way through literally trillions of descendant cells into a person. This astonishing feat requires tremendous co-ordination, with different cells coming on line at the right time and taking up their correct position. The complete instructions for all this are written in the DNA molecule, but for them to make any sense, there must be a timetable of which genes are read, and when. This schedule is set, to a large degree, by the proto-oncogenes.

The proto-oncogenes have changed the course of biology. They provide a deeper understanding of cancer than we have ever had before. When things are going well, the proto-oncogenes orchestrate the growth and development of the organism. They become active at certain times, and once they have performed their piece, they remain silent until called upon again. When this order is disturbed, everything falls apart, and the programme of co-ordinated growth cannot be sustained. It is a shambles.

An issue of paramount importance is how do the proto-oncogenes misfire to make a cell become malignant. Broadly speaking, there are two ways in which this can occur; either they are switched on at the wrong time or they undergo a mutation. For some time, biologists have observed that some types of human cancers are associated with certain chromosomal abnormalities. One piece of a chromosome breaks off and becomes attached to another. This causes genes to move from one chromosome to another, where they come under the influence of different regulatory switches. If it so happens that the switch is thrown to the on position, the transported genes also become active.

The importance of this reshuffling of genes becomes evident when the locations of the known proto-oncogenes are mapped to the recognised chromosomal defects in cancers. The trick is to find out if the proto-oncogenes are carried in the displaced fragments of chromosomes. A revealing example of this is seen in Burkitt's lymphoma, a particularly virulent malignant lymphoma. A frequent finding in this cancer is that a fragment of chromosome 8 containing a proto-oncogene, called *myc* oncogene, moves to chromosome 14, whose genes code for antibodies. The next piece of evidence is that the malignant cells of Burkitt's lymphoma are derived from lymphocytes that are continuously producing antibodies. By leaving its native chromosome, the *myc* proto-oncogene loses its usual regulatory switches, and its relocation on a different chromosome puts it in a foreign province, an active antibody-producing region. The *myc* proto-oncogene, bowing to a new authority, allows free access to its information. As its message gets around, the cell's everyday activity is subverted, and it responds by provoking endless cycles of cell division.

The relocation of the *myc* proto-oncogene to a "hot" spot in a different chromosome leads to its activation at the wrong time. Similar dislocations of other proto-oncogenes occur in some other human cancers, suggesting that this is an important means of the inadvertent activation of these genes.

In addition to the untimely activation of the proto-oncogenes, some undergo changes or mutations that set the cell on the track to malignancy. A mutation in the DNA causes a change in the amino acid make-up of the protein for which it codes. This can be severe enough to distort its three-dimensional structure and interfere with its function.

The *ras* proto-oncogene is a good example of the dire effects of a genetic mutation. A single mutation in one of the bases converts the normal proto-oncogene into a powerful inducer of cancer. In some cancers, a base in a codon GGC is changed to GTC. This causes the substitution of the amino acid valine for glycine in the *ras* protein. A small change, but strategic enough to alter its structure and function.

The *ras* protein is part of a system that transmits biological signals from the outside to the inside of the cell. When the cell receives a signal, it is transmitted chemically across the cell membrane through transitory changes in the *ras* protein complex. Once the message is sent, the system shuts down and returns to its ground state. When another message is received, the system is again fired, and the process is repeated.

A mutation in the *ras* protein causes the system to become trapped in an activated state, with a constant, unregulated transmission of signals. This deludes the cell, which interprets the uninterrupted barrage as a message for continuous growth. The cell moves, without pause, from one cycle of division to the next.

For a cell to remain in harmony with its neighbours, it must respond precisely to the inputs it receives from outside; any straying from the rules agreed to from birth would have harmful effects. Cells are interdependent, always tuned in to messages from others, and sending out their own. The scheme is a wonderful one, but it is not infallible. When it breaks down, the stage is set for the formation of a cancer.

In addition to the proto-oncogenes, another class of regulatory genes has been discovered. The disease that provided the most useful information about these is retinoblastoma, a rare eye cancer which usually develops in children by age five. There are two forms of the cancer; one is inherited, while the other is not. The children of those with the inherited form of retinoblastoma are also susceptible to the disease.

The inheritance of the *rb* gene, as it is called, is not, however, sufficient by itself to transform a normal retinal cell into a malignant one. A second mutation is needed to bring about the malignant change. The two mutations that are necessary before a retinoblastoma can develop are, in fact, the loss or deletion of the *rb* genes on the pair of chromosomes in the cell. Thus, unlike the proto-oncogenes which must be activated, the two *rb* genes must be silenced, suppressed.

At first, the *rb* gene was thought to be an oddity in an obscure cancer, but problems with the gene have now been found in several other common cancers, such as those of the breast, lung, and bladder. Since the description of the *rb* gene, at least ten other similar genes have been found in human cancers. A well-known gene of this genre is the *p53* gene, which is implicated in a variety of cancers. When it is mutated, the *p53* gene fails to perform its usual function of regulating the cell cycle, and thus allows the uncontrolled proliferation of cells. Collectively, these genes are called tumour suppressor genes, and they now appear to be as important as the oncogenes.

A different route to the development of cancer has recently been discovered. This is a defective repair mechanism for correcting errors in DNA. During the replication of DNA, a group of genes serve as proof-readers, making sure that the new DNA strand is faithfully copied. A mutation in one of them allows mistakes to slip through uncorrected; errors accumulate upon errors in the new DNA, and its message becomes muddled.

The genetic jigsaw puzzle is slowly being put together. A small number of regulatory genes supervise the growth and development of cells. Under normal conditions, they work together, ensuring that the organism performs the critical functions of making new cells and replacing old ones. The mechanism is quite a good one when used with precision, admirably designed to provide the checks and balances for orderly growth. But when it breaks down, the system wanders away from this orderly programme. Out of this, a cancer is born.

The genetics of cancer of the large bowel has been studied in some detail. Several tumour suppressor genes and at least one proto-oncogene are involved, accumulating as the disease progresses from a benign polyp in the bowel into an aggressive, malignant tumour that kills. Recently, the class of repair genes has also been found to be important in some of the heritable cancers of the bowel. A mutation in any one of these increases a person's susceptibility to such an extent that a careful programme of screening for cancer is essential for anyone with this genetic defect.

In breast cancer, a common cancer which affects one in nine women, multiple genetic mutations act in concert. At least three oncogenes are overexpressed, and nine tumour suppressor genes are deleted. The interaction among these various genes is complex. Add to these genetic defects the many known predisposing factors, and it becomes a real challenge to sort out. In the past few years, two repair genes have been identified which play an important part in about 10 percent of women with breast cancer; these are called the BRCA1 and BRCA2 genes. An inherited mutation in either increases a woman's susceptibility to breast cancer. Screening for these genes in women at high risk for the disease would in all likelihood open options for prevention.

CHAPTER 7

DESIGNER GENES

As we come to the close of the twentieth century, we can look back with considerable amazement and justifiable pride at the medical achievements since 1900. A good standard of health is so commonplace in the industrialised world that it goes largely unnoticed. We are living longer than ever before, and with each passing decade, our life expectancy continues to rise. It is sometimes difficult to understand claims to the contrary, warnings from some quarters that we are facing an impending catastrophe. Problems still face us, to be sure, and we do occasionally suffer the fall-out of our zeal for technology, but nowhere is a threat looming so ominous that we are rendered helpless, on the verge of extinction.

The history of modern medicine lies on a solid foundation — an understanding of the root cause of diseases. Until the rise of the biological sciences, much of what was done in medicine was empirical. The therapies down through the centuries reflected this, based, as they so often were, on the superstitions of the day. Much of the reason that these home-spun myths endured was the remarkable capacity of the human body to heal itself. The major part of the family-doctor's practice even today deals with illnesses that are self-limiting. The common cold is common, as is colic in babies. Give Nature a chance, and with just a little patience, in a few days everything will be just fine again. The body will respond to put everything right; health is restored. This is the built-in durability of the human organism. With such power to redress most everyday illnesses, just about anything can be touted as the definitive remedy: chicken soup, garlic, megadoses of vitamin C, or hot buttered Demerara rum.

In the mid-1800s, biology was put on a firm footing by Schwann and Schleiden with their description of the cell as the basic unit of living things. Rudolph Virchow grasped the significance of this seminal discovery and is credited with extending it to health and illness. "We can go no farther than the cell," he wrote. "It is the final and constantly present link in the great chain of mutually subordinated structures comprising the human body." He introduced a new way of looking at the body as a cell-state in which every cell was a citizen. Reflecting the spirit of his involvement in the German revolution of 1848, he envisioned the body as a "society of living cells, a tiny well-ordered state." In his view, a disease was a civil war between cells. What this implied was that a deep understanding of health and disease could not be possible without a deliberate and careful consideration of the cell.

Too often today, some seem in a hurry to denounce the advances of biomedical science, attributing all our misfortunes and ills to a runaway technology. They hanker for the good old days of simplistic treatments, relinquishing our claim to revel in the new age of science. They seem more comfortable with the old and tried, and betray a fear of the new and promising. How can we explain the truly extraordinary improvement of our lot but through the reasoned and sensible use of the tools of science and technology?

Unquestionably, a major medical advance this century was the discovery of antibiotics. It absolutely changed the world. Until it entered the scene, we fell easy prey to the virulent strains of germs around. Epidemics swept across the globe, leaving tens of thousands dead. At the turn of this century, tuberculosis and syphilis prematurely ended the lives of countless thousands. We hardly seem to remember those days when these perennial killers raged on. Yet, still among us today are some who in the mid-1940s owed their lives to a brand new kind of drug — penicillin. The memory is not so old afterall. Today, antibiotics are everywhere, and there is no denying the impact they have had on our health. It is but one example of the forward march of medicine.

The newest biotechnology, some ten years old, offers the same or greater potential to change our lives. We get a glimpse of the scope of this when the first human gene was cloned and put into human cells to correct a genetic defect in 1990. It is the final stretch of the journey that started in the mid-1800s when the cell was first described. And what a journey it has been! From that first observation of its existence, we have worked out the make-up of the cell, defined its complex biochemical reactions, and figured out its subtle regulatory systems. We are now ready to embark on the final venture: gene therapy. No longer are we onlookers but doers. We have the technology to potentially correct errant genes.

In 1987, the Human Gene Therapy Subcommittee of the National Institutes of Health in the US rejected a protocol to treat children born with a certain genetic disease, called severe combined immunodeficiency disease or SCID, with gene therapy. In SCID, the gene for an enzyme, adenosine deaminase or ADA, is missing in cells. These children suffer repeated infections, and are usually confined to sterile, artificial environments for their protection. Most die at an early age. The proposal before the Subcommittee was to replace the missing ADA gene in bone marrow cells, using a virus to ferry a cloned ADA gene into them.

The first protocol was turned back, but the debate was underway. After extensive review and careful clinical trials of the safety of gene transfer in terminal cancer patients, permission was eventually granted for gene therapy in children. On September 14, 1990, the first therapeutic human gene transfer took place at the National Institutes of Health. A young girl, aged four, with SCID was infused with her own blood lymphocytes that had been bioengineered in the laboratory. The missing ADA gene in the lymphocytes was replaced with a normal one, using a retrovirus to carry it into the cell where it was activated to produce the ADA enzyme. The function of the lymphocytes, which bore the brunt of the genetic misfortune of SCID, was successfully restored. Gene therapy came of age.

At present, the most efficient way to transfer a gene into a human cell is with a retrovirus. The virus has specific proteins on its surface that dock with matching receptors on the cell's surface. This allows it to gain entry into the cell where it releases its RNA genome. After this, the viral RNA is converted into a double-stranded DNA with the viral enzyme, reverse transcriptase. The process may be directed by the viral enzyme, but the raw materials for the new viral DNA, called a provirus, come from the cell. From the cytoplasm, the provirus migrates to the nucleus, where with another viral enzyme at its service, it insinuates itself into the cell's DNA. Here, it looks for all the world like any other gene. The provirus is now transcribed into numerous RNAs, just like any other cellular gene. In the next step, the viral RNA genomes become wrapped in individual protein shells to form new viruses. These then burst out of the cell to infect other nearby healthy cells.

It would seem a hazardous business to use retroviruses as vehicles for gene transfer, since they can cause human diseases, some extremely serious. The human immunodeficiency virus (HIV), for instance, belongs to this family of retroviruses. It takes some courage and much confidence in what we know about these agents to manipulate them to our own ends. The debate on just these points has been lively. The knowledge is already there, and the rage of the argument is about the application of the technology to treat patients.

To avoid the risk of illness, the retroviral genetic machinery is re-engineered. Those viral genes responsible for its replication in cells are removed, crippling the virus. It is still able to infect a cell, and its genome, now containing the human gene of interest, becomes inserted into the cell's DNA, but it cannot make new viruses. The immediate issue of safety and security of the transplantation of genes by viruses is assured.

In addition to retroviruses, other families of viruses are also being developed or considered for gene therapy. A promising group is the adenovirus. Unlike the retroviruses, their genome is a linear, double-stranded DNA molecule, and this makes them different.

Adenoviruses are common viruses found in all parts of the world. They usually infect the upper respiratory tract, and, in fact, the name comes from their initial identification in human adenoids. About 5 percent of acute respiratory diseases in young children are caused by adenoviruses. Typically, they cause cough, nasal congestion, headaches, and coryza. Occasionally, there are fevers, chills, and muscle aches. Adenoviruses are also responsible for the swimming-pool eye infections that tend to occur in outbreaks at children's summer camps. At times, the virus may cause gastro-enteritis, from which full recovery is the rule in just a few days.

In the infected cell, the viral DNA replicates in the nucleus into new viral DNA genomes. As this is going on, the viral genes are transcribed into messenger RNAs, from which viral proteins are made. These assemble into the viral protein shells. Next, the new viral DNA genomes enter the protein shells to form complete viruses. In the final step, the infected cell falls apart, releasing the new viruses, which infect neighbouring cells to keep the infection going.

The proclivity of the adenovirus to infect cells of the upper respiratory tract has been exploited in an experimental treatment for cystic fibrosis, a disease which affects about 30,000 persons in the US and 3,000 in Canada. It is a devastating genetic disease that claims its victims at a young age, few of whom live past 30 years. The gene responsible for cystic fibrosis was identified in 1989, and it provided a platform to launch this new treatment. The cystic fibrosis gene can be successfully transplanted into the cells of the respiratory tract with an engineered adenovirus, which is sprayed directly into the airways. The gene becomes active in the cells and replaces the malfunctioning gene. Even if only 5 to 10 percent of the airway cells pick up the virus, this would be sufficient to restore the genetic defect, and so prevent the accumulation of mucus in the lungs, the hallmark of cystic fibrosis. Unlike the retrovirus whose DNA becomes grafted, as a provirus, into the cell's DNA, the adenoviral DNA remains free in the nucleus, unattached to any chromosome. When the infected cell dies and is sloughed off, the viral DNA is also lost. Hence, repeated treatments may be required to control the disease.

A similar family of viruses, the parvovirus, is also of interest for gene therapy. Parvovirus causes the "fifth disease," a common childhood illness, often accompanied by a rash on the cheeks. The virus also occasionally affects adults, in whom it causes joint pains. Although most cases of parvoviral infections are minor, a more virulent strain, named B19, attacks young red blood cells, leading to serious complications in patients with certain chronic anaemias, such as sickle cell anaemia. Infrequently also, infection during pregnancy can result in foetal death.

Parvoviruses are the simplest DNA viruses to infect humans. Their genome is a short, single-stranded DNA molecule. The interesting feature of parvovirus is that its genome inserts itself at a specific site on chromosome 19. Thus, unlike the retrovirus whose genome can end up just about anywhere in any of the chromosomes, the parvovirus has a known destination. This offers a means to deliver a gene into a cell with predictable results.

The landmark experiment in children with SCID in September 1990 at the National Institutes of Health has prompted an earnest study of gene therapy. Diseases caused by inherited mutant genes are the obvious choice for gene therapy. In addition to SCID and cystic fibrosis, another disease of interest is familial hypercholesterolaemia, in which the genes for the cell receptors for low density lipoprotein, or LDL, are defective. This leads to an extremely high level of LDL cholesterol in the blood, causing premature atherosclerosis and early heart attacks, sometimes even in children. An experiment to treat this deadly condition with gene therapy is in progress. Liver cells from affected persons are removed surgically and treated in the laboratory with retroviruses, modified to carry the normal human LDL receptor gene. The treated cells are then injected back into the liver's blood supply, whence they settle in that organ and grow. Once these bioengineered cells start to function, the new gene is expressed, and the blood cholesterol level falls.

The roster of diseases for which gene therapy is under investigation continues to grow. Among them are blood disorders like thalassaemia, sickle cell anaemia, and haemophilia; inherited emphysema; diabetes; and Duchenne's muscular dystrophy. Gene therapy has a vast potential. At present, about 4,000 genetic disorders are known, and some 2,000 responsible genes have been identified, all of which can theoretically be fixed by this new method of treatment. Indeed, by 1996, at least seventy protocols have been started to investigate the value of gene therapy in the treatment of human diseases.

Gene therapy is also being vigorously tested as a treatment for cancer. Three major methods of cancer therapy have been developed: surgery, radiation therapy, and chemotherapy. Varying success has been observed with immunotherapy, and its role in the overall management of cancer remains marginal. About one-half of all persons with cancer are cured outright in the 1990s. This figure does not include the most common cancer, skin cancer, in which an overwhelming majority of persons are cured. If these are included, then the cure rates are even more impressive. Unfortunately, however, results of cancer treatment seem to have reached a plateau. Every inch of new ground has to be fought for. Some cancers cannot be removed by surgery, some survive high-dose radiation treatment, and others remain obstinate to the best chemotherapeutic drugs presently available. To overcome the problem of the remaining one-half of cancers, unbudged by our present treatments, we may well have to look at new approaches.

The first attempt at cancer gene therapy was to infiltrate the cancer with blood lymphocytes that were loaded with a toxic molecule. That molecule was tumour necrosis factor or TNF. TNF was first discovered in 1893 by William Coley, a surgeon at the New York Cancer Center. He observed that cancers in few patients shrank after a bout of erysipelas, a Streptococcal skin infection. Later, at the Memorial Sloan Kettering Cancer Center in New York, L.J. Old found out that this was caused by a substance which he named tumour necrosis factor.

Our body normally produces TNF in response to an infection by bacteria or infestation by some parasites. The amount made is extremely minute, just sufficient to deal with the infection. Large amounts evoke a violent and often lethal reaction. If overwhelmed by large amounts of TNF, as by an injection through a vein, the body reads this as the worst of bad news. There is a fever, the blood pressure falls, and, soon, shock ensues.

What was sought was a means to deliver TNF to the cancer only, thus avoiding the exaggerated reaction to an intravenous injection. A clever way has been devised to do just this by using a special type of blood lymphocytes, tumour-infiltrating lymphocytes, which are usually mobilised against cancer cells. They travel through the blood stream and assemble in the cancerous tissues, where they engage in a showdown with the alien malignant cells. It appears that in most clinical cancers which have already grown to a detectable size, the battle is already lost, and the tumour-infiltrating lymphocytes are outgunned.

Steven Rosenberg and his colleagues at the National Institutes of Health inserted the gene for TNF into the tumour-infiltrating lymphocytes with a retrovirus. The genetically engineered lymphocytes were then injected into the blood, from where they infiltrated the cancer, releasing a large amount of TNF. The hope was that by planting these molecular explosives within the cancer, it would fall to pieces. The approach has had some impact in the treatment of some cancers, like malignant melanoma, but its overall application to cancer therapy appears limited. Nonetheless, it has taken cancer therapy in a new direction by bringing the powerful tools of molecular biology to the bedside.

The engineering of tumour-infiltrating lymphocytes was an indirect approach to control cancer. A more frontal assault can now be contemplated. The genes responsible for many cancer are known, and with this comes the expectation of thwarting the relentless march of the disease. An example of this new approach is an experimental protocol to treat lung cancer.

Lung cancer is the major cancer killer in western countries. Unfortunately, most cases are diagnosed at a stage when complete surgical removal is not possible, or even when this is feasible, relapses are common. To deal with any residual disease after surgery, doctors are injecting retroviral vectors with a normal gene to replace the defective one in the cancer. In those cancers with a mutant *p53* tumour suppressor gene, a normal gene is sent into the malignant cells to replace it, and so restore normal cellular function. In this way, the proliferative thrust of the cancer cells is blunted, and they are reined in.

Another approach is to prevent the expression of the mutant *ras* gene in lung cancer cells. In normal cells, only one strand of the double-stranded DNA is transcribed into messenger RNA. This is the "sense" strand; the other, logically, is the "antisense" strand. A gene can be crafted in which the antisense strand is transcribed into messenger RNA. This sticks to the messenger RNA of the sense strand — observing the universal rule of base pairing between A and U, and G and C — and neutralises it. As a consequence, the message of the sense strand is lost, and it is not translated into a protein.

The difference between the normal and mutant *ras* genes, although small, is critical enough to make this strategy potentially useful. A gene that expresses an antisense RNA for the mutant *ras* gene could be delivered into the cancer cells by a retrovirus. Here, the antisense RNA binds to the mutant *ras* messenger RNA and prevents its translation into a protein. The mutant *ras* gene is subverted, and the wayward cells are brought under control. As the antisense RNA for the mutant *ras* gene does not bind to the normal *ras* messenger RNA, cells with the normal *ras* gene remain unaffected.

There seems no limit to the imagination of the new breed of molecular geneticists. The clever method to treat brain tumour with gene therapy is a good example of this. One of the problems faced in using a retrovirus to transfer genes into cells is that only a fraction of them actually take up the virus. To infect all the cells by injecting the virus directly into the cancer is not easy or, for that matter, even possible. But what about producing new retroviruses right in the cancer?

Brain tumours seem a likely candidate for this approach. By injecting cultured cells, which are routinely used to grow retroviruses in the laboratory, into a brain tumour, they are made on site. These new retroviruses infect the nearby cells, and, over a period of days, many more come to harbour them. The success of this strategy lies in the fact that the normal, mature brain cells do not usually divide, and, therefore, do not admit the provirus into their genome; they stand by, silently, protected. The cells of a brain tumour, on the other hand, are actively dividing, providing the opportunity for the provirus to edge its way into their chromosomes.

The genome of the retrovirus used in this experimental treatment contains a new gene which codes for a special viral protein. When the provirus is taken up and that gene expressed, the malignant cell becomes labelled with a new, acquired trait. This distinguishes it from the normal brain cell, and also makes it an easy target for certain antiviral drugs. A few days after injecting the cultured cells with the retrovirus into a brain tumour, an antiviral drug is given intravenously to the patient. The cancer cells, which now present a new face, are recognised by the drug and are killed off. Clinical trials of this novel way to treat brain tumours are in progress.

The technique of gene therapy may also be used to combat HIV-1 infection, whose control has so far proved a challenge. HIV multiplies in a subset of blood cells, helper T-lymphocytes, in which it undergoes many rounds of replication. The virus remains undetected for many years before it causes any symptoms. Eventually, however, the immune system breaks down, and overt AIDS emerges. With few effective antiviral drugs available, attention is turned to innovative treatment strategies.

The life cycle of HIV, like other retroviruses, is known, and this allows the design of new ways to interfere and block the process from infection to replication. A few different approaches are under consideration. First, when the retrovirus delivers its RNA genome into the cell, it is naked, exposed, in the cell cytoplasm. Here, it is vulnerable to enzymes, such as ribozymes, that cut RNA molecules into tiny bits.

Ribozymes can be devised that recognise the viral RNAs and slice them at several sites, stopping the virus in its track before it gets a foothold.

Another way to tackle the problem is the ultimate irony: using HIV itself as a weapon of its own destruction. In this, the genes needed for replication are removed from the HIV genome and replaced with other genes. These alien genes are then delivered into the HIV-infected cells, an ideal situation as the viruses have a distinct predilection for the T-lymphocytes. A gene that produces an antisense RNA to the HIV RNA is an example of how the tables can be turned on the virus. The antisense RNA binds to the viral RNA and knocks out its message.

Gene therapy is not without some potential risks. The retroviral vector brings into cells its provirus, which then becomes part of the cellular DNA. If it so happens that the provirus lands next to a normal proto-oncogene, it could inadvertently switch it on and set the cell up for pernicious changes by disturbing its normal pattern of growth.

So far, the technique of gene therapy is restricted to those with major genetic disorders or other serious diseases, all of which are fatal if untreated. The possibility that the germ cells, the egg and sperm, can be infected by the retrovirus seems remote, but it's a risk to be cognisant of. The implication of accidentally modifying the germ cells is that the new genes can be passed from parent to offspring, with, as yet, unknown consequences.

A third and, perhaps, the most immediate concern is the potential reactivation of the viral vectors. All viruses used for gene therapy are incompetent; that is, they cannot replicate, because their genome is greatly altered. A modified HIV, for example, is unable to go through all the stages of its life cycle in the infected cell. It contains only the genetic material necessary to regulate the alien gene it carries; all its other genes have been removed. But what if a cell is infected with a normal HIV and an engineered one? An exchange or sharing of genes between the virulent HIV and the modified virus could restore and might even enhance its virulence with detrimental results.

While the most effective way to deliver new genes into cells relies on this ancient invention of Nature, the virus, a second generation of synthetic substances is already under investigation. Wrapping the gene in a coat made of lipids and proteins could allow entry into cells, and this technique is now being developed. If successful, it would circumvent some of the risks associated with the use of live viruses.

The dream of doctors is to be able one day soon to take a vial off the pharmacy shelf and inject its contents, new genes or gene products, into patients with any of the large number of genetic diseases. These would get into the diseased cells and, by replacing a faulty gene with a good one, restore health. Today, this is still a dream. Nonetheless, with what we have witnessed in the past decade, it would be unkind not to give credit to human ingenuity. Not so long ago, few would have taken the idea of repairing and replacing defective genes seriously. However, the new biology has placed at our disposal the tools of genetic engineering possessing a power previously unimaginable.

The era of gene therapy is here today. Our good health is its promise for tomorrow.

CHAPTER 8

ORDER FOR FREE

Life on Earth is about 3.8 billion years old. For most of this time, simple one-celled organisms swarmed over the surface, occupying every niche. The transition from this mind-numbing simplicity to the rich complexity of those creatures made of many different kinds of cells took place only 600 million years ago, almost overnight, in the Cambrian. In one incredible swoop, life changed, and continues to change even today. The descendants of the original occupants are still here, but now they share their quarters with myriad other larger organisms, which roam over the surface, burrow underground, or swim in the deep seas. Over the time since the Cambrian, multicellular life has become increasingly diverse and complex. The upward trend has occurred by fits and starts, punctuated by bursts of innovations and quelled by mass extinctions. There is a continual turnover as new species replace existing ones, but life has maintained a progressively complex pattern.

Almost all multicellular animals and plants start life as a single cell, typically a fertilised egg, from which a large number of descendant cells are spawned. A human, for example, is made of more than fifty trillion cells of at least 250 different types. Each of these must occupy and function in its proper place. It is a tremendous enterprise, driven by the exchange of information among cells. Cells constantly receive and process information, and send out their own messages at the same time. How they react and interact are governed by their genes.

For some time, biologists have realised that the different cells arise depending on which genes are activated, and when. Cells multiply and move around the embryo, changing and eventually coming together into specialised tissues. How this occurs, however, is a mystery, since all the cells in an organism have exactly the same set of genes. Each cell division is preceded by a mitosis, during which the DNA is replicated and the full complement of chromosomes is passed on to each of the two daughter cells. What this implies is that the developing organism must switch some genes on, and others off, at prescribed times.

By turning to the right page each time, the organism choreographs the activity of its genes so that the right cell is made by reading the correct passage from the large book of information encoded in its DNA. Read a different passage, and another kind of cell is made. This is the magic of development; it allows a single cell to unfold into a full organism — a panda, a penguin, a person.

The co-ordination of this expression of genes is a crucial issue in modern biology. It is also a challenge, as cells contain a very large number of genes. Which of these becomes active at any time is almost certainly under the control of regulatory mechanisms that involve other genes. Understanding the genomic regulatory system is not easy, but an outline of just how such a system operates was deduced some time ago.

When embryologists cut pieces of tissues from one part of a frog's embryo and put them at another site, the transplanted tissue would sometimes grow as it would in its original site. For example, if cells from a region destined to develop into a leg were transplanted to another site, the adult frog would grow up with a leg in the wrong place. Repeat the experiment at an earlier stage in development, and the transplanted tissue developed appropriately in its new site to produce a normal frog.

This means that early in development, cells are versatile, able to switch from one form to another, but, later on, they become fixed. At some point, a signal must be sent to the cells that affects how their genes are expressed, what cells they become.

The first insight into how genes were regulated was gained in 1960 through the important work of the French biologists, François Jacob and Jacques Monod. They carried out a series of experiments on the common gut-dwelling bacterium, *E.coli*, from which they derived some general principles of gene regulation. These are also applicable to the cells of higher plants and animals.

When the milk sugar lactose is available, *E.coli* immediately starts to make the enzymes to burn it for energy. Jacob and Monod discovered that the genes for these enzymes are switched on and off depending on whether lactose is around or not. They had a hunch how this might take place. They suggested that the *lac* genes, as they are named, are active only when lactose is available. Otherwise, a regulatory protein prevents their transcription by binding to the DNA next to them. However, when lactose is present, it dislodges this protein from DNA, and transcription of the *lac* genes could then take place. This turns out to be exactly what happened, and for their brilliant work on the regulation of genes, Jacob and Monod were recognised by the Nobel Prize committee.

The existence of feedback mechanisms by which the bacterial genes are switched on and off helped in clarifying the details of how genes are regulated. For instance, *E.coli* has three structural *lac* genes, called *lac* ZYA, to use lactose. They allow it to absorb lactose from the outside across the cell wall into the cytoplasm, where it is burned for energy. The ZYA genes are under the control of two elements, referred to as the promoter and operator, located just upstream from them on the chromosome. It does not end there. The operator is, in turn, regulated by a protein, while the promoter is itself regulated by four molecules. When there is no lactose around, the promoter and operator are inactive, and they shut down the *lac* ZYA genes. Add lactose, and they become activated, turning on the *lac* ZYA genes. Unless all the right conditions are met, transcription of these genes does not occur.

It may seem an overwhelmingly complicated matter for bacteria to take up and use lactose as an energy source. Bacteria usually get their energy by burning glucose and will use this preferentially to lactose. Only when glucose is in scant supply, will they turn to lactose. We find an explanation for this fastidious behaviour in their use of lactose by their lifestyle. Bacteria are independent, free-living organisms which must survive in an environment that is always changing and where a supply of food is never assured. They must, therefore, be able to efficiently use any food present, and to swiftly respond to any fluctuations in its supply; their survival depends upon it.

The bacterium appears to have struck a workable compromise. It might use most of its genes during its short life, but it switches them on or off rapidly in response to its immediate needs. It produces no *lac* enzymes when lactose is absent, but once it gets a whiff of it, it responds unhesitatingly by producing them to exploit its availability. Within minutes of lactose appearing in its environment, *E.coli* produces *lac* ZYA enzymes, reaching up to 5,000 molecules in each cell. As soon as lactose disappears, the production of the enzymes stops almost as quickly as it originally started.

This allows the bacterium to make use of lactose when it is available, but avoids expending time and energy in unnecessarily producing the enzymes when there is none around. There is no need to produce them all the time; the system is finely tuned to react to the prevailing conditions. Unless it is metabolically agile enough to respond in this fashion, the bacterium will lose out in the fiercely competitive world where billions of other organisms are forever jostling for better advantage.

The component that facilitates this rapid reaction is the regulatory proteins. In the absence of lactose, they ensure that the adjacent structural genes are not transcribed needlessly. But when lactose is present, they free the genes, and production of the enzymes follows.

An important distinction can now be made between two classes of genes. First, there are the structural genes that code for proteins, which are used for various cell parts or serve as enzymes. Second, there are genes that control the activity of the structural genes; these are the regulatory genes. In some cases, they prevent the transcription of a structural gene; in other cases, they enhance its transcription.

At this point it is well worth asking: How is regulation achieved in the genomic systems of higher mammals that are far more complex than *E.coli*? The first obvious level of complexity is the sheer number of genes involved. The human genome has enough DNA to code for about 2,000,000 average-sized proteins. This, however, is almost certainly an over-estimate, since a great portion of the DNA in the genome may not be directly transcribed into proteins. A good estimate of the number of structural genes in a human cell is 100,000, and if we assume at least one regulatory gene for each structural gene, then the total number of genes might be 200,000. This is probably a conservative estimate, but the important point is that a large number of genes are present in the cell.

In the cells of higher organisms, not all the genes are active; in fact, most are quiescent, and only those required for making each cell what it is are actually switched on. In all cells, however, a common core of genes is present, often accounting for up to 70 percent of all the active genes. An immediate explanation for this is that these common genes are needed for conducting general cell metabolism, the workaday details of living. This leads obviously to the conclusion that only a small proportion of the total number of active genes in a cell is responsible for the differences between it and a cell of another kind. A liver cell, for example, shares a large set of common genes with a nerve cell, whereas the different genes that make each distinct, one from the other, are small in number, indeed.

This explanation, which at first seems attractive, may not, however, be true. In *E.coli* and other bacteria, there are only about 2,000 to 3,000 structural genes that subserve the regular housekeeping details. Yet, they live in an environment which is always changing and to which they must respond promptly in order to survive. On the other hand, the individual cells in more complex multicellular organisms are much less buffeted by outside changes; they live in more stable conditions.

It would, therefore, seem strange that the cells of higher organisms require considerably more genes to conduct their everyday duties than free-living bacterial cells. Most of these genes may, in fact, be responsible for functions apart from the daily routines of cell life. One possibility is that they regulate the activities of other genes. While it may be true that not many genes distinguish a liver cell from a nerve cell, it takes a co-ordinated genetic system to ensure that only those that go into making a cell one kind or another are brought into service.

The regulatory mechanisms can be exquisitely sensitive. A particular cell may use only a small proportion of all its genes, but selecting the right ones is not simple. Each cell must know which genes to switch on, and when. Seen in this light, gene regulation is a paramount property of the genomic system of higher organisms.

How would a genomic system with hundreds of thousands of genes assemble itself into a working whole?

A conceptual model of the genetic regulatory system, embraced by most biologists, has been a hierarchical one. A few master genes at the top command lower echelons of genes. Orders issued at the top are passed down, in a direct manner, through the ranks. At the lowest ranks are the structural genes, diligently carrying out their specific duties, maintaining the integrity of the whole organism.

This scheme is a simple one, but is it accurate? Its major flaw is that even a minimal error at the higher ranks can unhinge the entire system by magnifying its dire effects as it cascades downstream. Avalanches of changes, possibly incompatible with survival, spread widely through the genetic system.

A more plausible scheme is one of a regulatory network rich in feedback loops. Here, orders issued by high ranking genes pass through intermediate genes, which process information coming in from various other sources. How these respond depends on the mix of signals they receive. This allows for a constant interplay amongst the genes; they are part of a genetic circuitry, in which the switches can be turned on and off.

It is pertinent to ask how such a genetic wiring system, which allows the efficient and integrated relay of messages, could arise. Stuart Kauffman, a theoretical biologist from the Santa Fe Institute in New Mexico, has been influential in shaping our thinking about this. Consider a network of N structural genes. To remain in touch with the other genes in the system, each gene must receive at least one input from another gene. In fact, the evidence points to each gene receiving more than one input. Some interesting patterns arise as the number of regulatory connections, M, increases (Figure 15). When M is less than N, there are sparse connections between the genes, and for all intents the system remains poorly connected. As M increases and approaches the value of the number N, however, small, isolated, branched circuits begin to form. When M equals the number of structural genes N, an interesting pattern is observed. At this point, interconnected loops emerge abruptly. As M becomes greater than N, the system becomes richer in feedback loops. The crucial point is how very different this scheme of genetic organisation is from the simple hierarchical command structure.

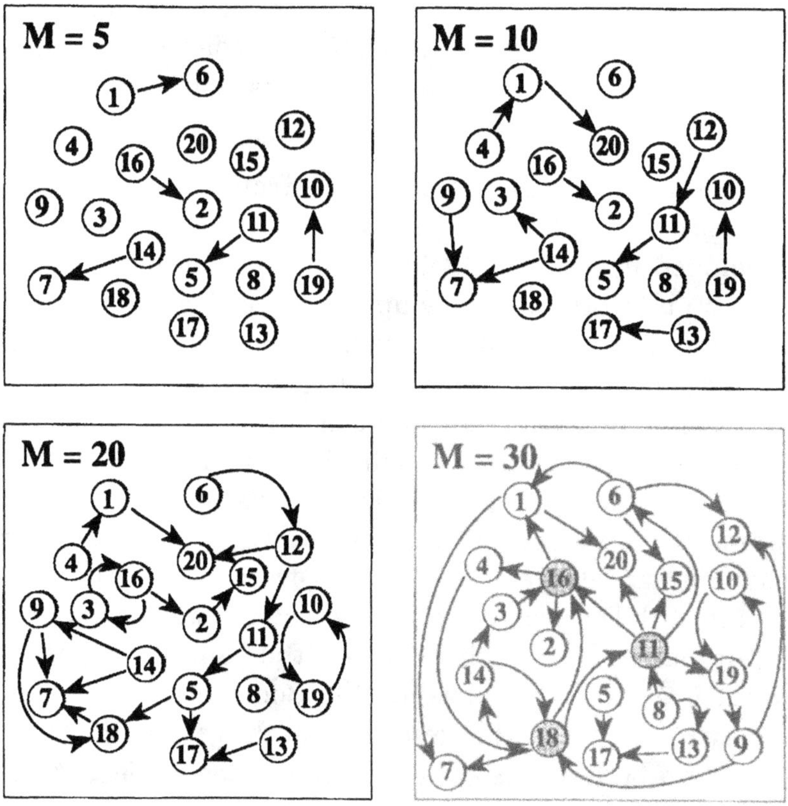

Figure 15. Kauffman's model of interconnectedness among genes. A fixed number of genes, shown as circles, are connected by an increasing number of arrows.

Kauffman studied the behaviour of a genetic network, in which each gene is either active or inactive, depending on the inputs it receives from other genes. From this, he gained an insight into how the genetic circuitry is scrambled into a workable whole. Clearly, some rules must be built in for the genes to respond in a certain manner. Provide them with a particular set of conditions, and specific genes are switched on. Change the conditions, and different genes are switched on. This predictable response suggests that there may well be rules that dictate how genes behave.

The rules of the genetic circuitry come from Boolean functions, a system of mathematical logic developed by George Boole, an English mathematician. Boolean networks are made up of several elements, each of which is either on or off at any given moment. Whether an element is on or off — active or inactive — depends on the inputs it receives. The system is continually updating itself. At each moment, each element in the network examines the signals coming in from its links with other elements, and becomes active or inactive. How the incoming signals are interpreted is governed by preset rules (Figure 16). For example, a gene which is connected to two inputs might be switched on if either one or the other input is active. This is called the Boolean OR function. Alternatively, we might prescribe that the gene be switched on only if both inputs are active; this is the Boolean AND function; it requires both inputs to be active.

We can look at the operation of a hypothetical genetic network. Consider a simple network of three genes, a, b, and c, each of which receives input from the other two (Figure 17). We can prescribe the particular Boolean rule: Gene a is governed by the AND function; that is, a is active only if b and c are active. We can also prescribe for

Boolean OR Function

| Input | | State of gene receiving input |
Gene 1	Gene 2	
Active	Active	Active
Active	Inactive	Active
Inactive	Active	Active
Inactive	Inactive	Inactive

Boolean AND Function

| Input | | State of gene receiving input |
Gene 1	Gene 2	
Active	Active	Active
Inactive	Active	Inactive
Active	Inactive	Inactive
Inactive	Inactive	Inactive

Figure 16. Boolean Functions, showing the state of a gene receiving inputs from two other genes.

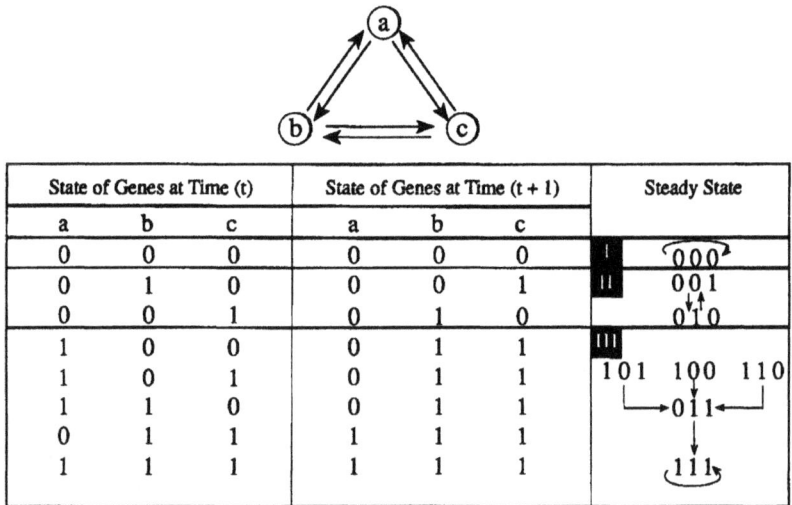

State of Genes at Time (t)			State of Genes at Time (t + 1)			Steady State	
a	b	c	a	b	c		
0	0	0	0	0	0	I	000
0	1	0	0	0	1	II	001
0	0	1	0	1	0		010
1	0	0	0	1	1	III	
1	0	1	0	1	1		101 100 110
1	1	0	0	1	1		011
0	1	1	1	1	1		
1	1	1	1	1	1		111

Figure 17. A simple Boolean network of three genes, each an input to the other two. Adapted from Kauffman SA. Antichaos and adaptation. Copyright August 1991 by Scientific American. All rights reserved.

the sake of the exercise that genes *b* and *c* are governed by the OR function; that is, an active input from either of the other two genes is sufficient to activate it. At a given instant, each gene examines the inputs from the other two, and immediately adjusts its own state of activity as determined by the preset Boolean rules.

The network proceeds through a series of so-called states. The individual genes respond to the incoming signals according to the rules, and become active or inactive. The network then proceeds to the next state, and the process repeats itself. In the scenario of two inputs per gene, the number of possible different states of the network of three genes is eight, as is depicted in Figure 17.

Under certain conditions, a network may pass through each possible state before repeating any of them. Most commonly, however, the network settles in a few states through which it cycles repeatedly. This repeated series of states is known as a state cycle. In our example, the system eventually hits a series of three states around which it cycles repeatedly.

The number of possible states even in small connected networks increases rapidly in a geometric manner. For more than eight genes, it is next to impossible to calculate this by hand, and a computer would be necessary. For a network with a hundred genes, each with two inputs from other genes, there would be some 10^{30} possible states; for 100,000 genes, the number staggers the imagination.

It would seem at first glance that to expect any order in a genetic network made of 100,000 genes is impossible. Kauffman analysed the behaviour of a network made up of N genes, each receiving K inputs, where N and K can be any number. At a specific point, and somewhat unexpectedly, he came across a stunning observation. Networks made up of a vast number of genes, even in the hundreds of thousands, can under certain specific conditions settle down and behave in an organised fashion.

To understand how Boolean networks can arrive at a state of order, we should first examine them more closely. At one extreme, we can imagine a network where each gene is directly linked to every other gene, dubbed K = N networks. There is only a single possible layout for the network; once set, it is fixed. In this state, flipping even one of the input to any gene from on to off, or vice versa, reverberates through the entire network, and forces the other genes to realign themselves before stability is restored. The important point here is that the system settles in a different pattern after any perturbation, however slight. Such a system is maximally unstable, chaotic. The smallest of disturbance causes a major change in the pattern of the entire network almost immediately.

At the other extreme, each gene is linked to only one other gene in the network, termed $K = 1$ networks. Nothing interesting happens here; the system is simply frozen. Changing an input from, say, off to on causes a minute, local effect, but the remaining network remains unaffected and unchanged

These two extremes of maximal instability for $K = N$ networks and frozen rigidity for $K = 1$ networks are of no biological interest. A genetic network designed along either pattern would simply be unworkable; life cannot be sustained in these conditions. In between, however, some curious results are observed. With five or more connections for each gene, the system remains disordered in that any minor change affects the entire network in drastic ways. But as K falls below four and approaches two, something very fascinating happens. There is a dramatic change in the behaviour of the network. With only two inputs to each gene, there is the unexpected and spontaneous emergence of order. The network becomes stable, and even when disturbed, it usually settles back to its original state — quite an important property for living cells.

The discovery of an inherent simplicity amid highly complicated and complex systems is often a turning point in science. This does not happen often, but there is no denying its influence on the course of history. We have only to think of Mendel, Newton, or Einstein, to name a few, to appreciate this. The finding of an underlying simplicity in the organisation of our genomic regulatory system may well prove another such event in biology, something to celebrate.

There is a deeper significance to a network with two connections for each gene, the so-called $K = 2$ networks. The number of state cycles into which such systems settle is small. For instance, a network of 100 genes has about 8 state cycles; a 1,000-gene network has 33; and a 100,000-gene network, which is about the number of genes in a human cell, has 370 state cycles. This turns out to be approximately the square-root of the total number of genes in the network.

Earlier, in Chapter 4, we saw that for a wide range of species from different phyla, the number of genes increases in a definite way in relation to the number of cells: the DNA content doubles for each new kind of cell in a multicellular organism; in other words, the number of cells is roughly the square-root of the total DNA. It is of more than passing interest that the number of cells in various species, based on the total genes, and the number of state cycles in $K = 2$ networks, follow a similar mathematical progression. The parallelism is hard to ignore. The fit is not perfect, but biologically plausible. This observation speaks strongly in favour of a fundamental principle that underlies the genetic organisation of living things.

Kauffman referred to this crystallisation of order out of massively disordered systems as "order for free." For three centuries now, scientists have uncovered many of the secrets of the cosmos, armed with the mathematics of Newton and Einstein. The laws of the physical sciences are well defined; they are repeatable and they are predictable. The biological sciences have not been definable with such precision. The unpredictability and variability of living things seem to defy mathematical analysis.

Are we now, however, beginning to see the faint outlines of the laws that govern biology?

CHAPTER 9

READING THE MIND OF GOD

In 1988, a concerted effort to understand the human genome by mapping the base sequence of DNA took shape. Just considering the sheer number of bases involved in this task is daunting. In germ cells, there are three billion base pairs; when we take into account that there are four different bases in the genetic alphabet — A, G, T, and C — the amount of information is an astounding 750 megabytes. To establish the identity and correct sequence of these bases is an immense project, an undertaking of unparalleled magnitude in biology. It led Nobel Laureate, Walter Gilbert, a pioneer of the methods to study DNA, to remark that defining the entire human DNA is like pursuing the holy grail.

Human cells contain two copies of the DNA sequence, one from the mother and one from the father. The germ cells — the female egg and the male sperm — contain one copy, a unique mosaic of the two genomes from which it is derived. Until now, the study of human genetics was limited to observation of individuals, families, and populations. But this is changing. The powerful tools of molecular biology now make it possible to construct a detailed physical and genetic map of the entire genome. With this comes an unprecedented opportunity to look into the very heart of life. We have already seen how this provides biologists with a road map to venture into new territories.

How far we can go on this adventure?

The potential of genetic engineering is vast. DNA can be manipulated with ease, bits snipped off and stitched together, at will, to make novel molecules. We can send these into cells, where they become permanent lodgers and have a say about how the place is run. Questions about the merits of this enterprise might still be debated, but the new technology tackles diseases, like intractable cancers or genetic disorders, at their very root. Herein lies its power.

Perhaps, the most unexpected finding of modern biology is how fluid the genome is. Genes change in the normal course of their lives. Far more than just storing in a single-minded way the collective experience of our evolutionary history, genes are surprisingly flexible. A cogent example is the making of the millions of antibody molecules. Nature invented the mechanism to cut and paste DNA long before we stumbled upon it. There is nothing new under the sun.

The complex nature of the human genomic architecture raises the question of whether we can really understand life. The details are not straightforward, even when reduced to a catalogue of bases. Reading out strings of bases in DNA is meaningless. A small proportion of the genome, perhaps only about 10 percent, codes for proteins, and we can make some sense of this. The remainder, almost 90 percent, remains a mystery; it appears to co-ordinate all the cell's activities, but we now know little about this.

There are two different and distinct levels of operation of the genome. First, DNA is read and copied into messenger RNA, which, in turn, is read by the ribosome to make protein. The details of this functional level have been worked out and understood for some time. The second level of operation is more subtle and less well known: it is a central regulatory body. From here, instructions are sent out when to make proteins, which ones to make, and how much. Without such central control, nothing would go right; there would be no integration of the many parts, and the whole system would simply fall apart.

It now appears that regulation is achieved by special proteins specified by the DNA. These fit snugly into the DNA molecule on either side of the regions that code for RNAs, stopping or enhancing the transcription of the DNA message. Production of these regulatory proteins is, in turn, controlled by other proteins, and an elaborate meshed network of feedback loops forms.

There may well be certain laws of Nature that govern the development of the genetic network of living things, but defining them has been elusive. Ever since the discovery of the fantastic molecule of DNA, we have regarded it as the source of all order. We have come to accept that DNA carries all the necessary information to build and sustain every living organism, and even some viruses that are not even living in the sense of being able to conduct their own metabolism.

We think of DNA as a code or, in the new computer age, a programme that contains the genetic information to make an organism — a protozoan, a plant, a person. The information is passed on faithfully from one generation to the next. Equipped with the enormous bits of information in DNA, the organism goes about the business of reading it, and through the pattern of the bases in DNA sees which parts govern what. Provide it with the programme, and the organism takes over and does the rest.

Does it?

The idea fires the imagination. We have in our hands very ancient DNA. In 1991, Edward Golenberg extracted DNA from fossil magnolia leaves, some eighteen million years old. Not to be outdone, Rob de Salle recovered DNA from a 32-million-year-old termite trapped in amber. Amber is hardened, clear tree resin which becomes a sealed tomb, keeping water out and preserving insects for millions or even hundreds of millions of years. By 1993, the date was pushed back even further, out to 120 million, by researchers from California Polytechnic State University with the extraction of DNA from a very, very old insect. The intriguing question that this endeavour raises is, of course, can we reconstruct these long extinct creatures? With the complete genome, can we build, or more accurately rebuild, the organism?

The fundamental assumption here is that the link between the DNA programme and the organism is direct. In other words, an organism's DNA contains everything for building it. This has been the accepted view.

The idea of a genetic programme first found a home at the turn of the century in Weismann's doctrine, in which he argued that the germ plasm was passed in continuity from parent to offspring. In each offspring, the germ cell controlled the development of the organism, but was not itself part of the body. According to Weismann, the problem of inheritance was not one of how the structure of the parent was passed to the offspring, but was about growth and development. The germ cells contained a specific substance, the germ plasm, that had the "power of developing into a complex organism." That specific substance we now know as DNA. Weismann considered that germ cells were derived from the parents' germ cells, which in turn came from their parents, and so on, in perpetual continuity from the first origin of life. This view is implicitly held by biologists today, who continue to think in terms of a genetic blueprint, or programme, that controls development.

A boost to this way of thinking came in 1995 from Raul Cano and his team at California State Polytechnic University. They succeeded in recovering bacterial spores from a Dominican stingless bee trapped in 25-million-year-old amber. When they find themselves in unfavourable environments, certain bacteria can survive by enclosing themselves in a thick protective shell or spore, in which they remain dormant for an extremely long time. In contact with food, the bacteria can come out of their shells and start growing again.

Cano carefully retrieved the bacterial spores from the ancient bees and brought the entire organism back to life by providing the suitable growing conditions. The report of this extraordinary feat still needs verification by others, but if it holds up, it will be a landmark event.

There is a risk of taking the concept of a DNA blueprint for life too far. The genetic organisation of higher organisms that stands before us is puzzling in its vast complexity. There is no debate of the central role of DNA in the passage of genetic information from parent to offspring. However, the idea of DNA as a straightforward set of instructions sits uneasily with the observation that the early development of an embryo in multicellular organisms is not controlled by its own DNA, but rather by maternal RNAs and proteins stored in the egg. They travel down from egg to new-born, always there, passengers across generations.

Take the natural nucleus out of the fertilised egg and replace it with that of another species, and development continues along the same path as it would had the original nucleus been present. Only after several rounds of cell division, when the embryo's ground plan is sufficiently laid down, does each of its cells call upon its own genes to direct further development.

This departs from the conventional view of the fertilised egg as a distinct entity, fully capable of carrying out its own growth and development. It does bring a supply of nutrients and fuel to burn for energy from its mother, but the observation that early embryonic development remains under the powerful hold of maternal molecules of the preceding generation raises some new questions.

It may seem a small point, but the implications are far-reaching. The first rounds of cell division are under the influence of the maternal genes before the embryo's own genes take over. What last minute instructions are given to these early cells by the mother that would shape their destiny? What secret does the mother whisper to her new embryo at this last moment before it takes leave and sets out on its own journey?

The concerns of Generation X aside, there is at the molecular level no generation gap. We accept the continuity of DNA across the generations; in fact, however, the connection may be stronger and more vital than we think. Specialised molecules are pre-packaged in specific areas of the egg. As it divides, different daughter cells come to contain different subsets of critical molecules. This sets the stage for the development of different kinds of cells, the prerequisite for multicellular life. Under the watchful eye of maternal molecules, the process gets started. The crucial point is that the DNA in the embryo does not initiate development.

There is another curious observation. The proteins of chimpanzees and humans are nearly identical, meaning of course that the DNA base sequences are also very similar. Nevertheless, the differences between the chimps and us are great, at least as they appear to the human observers, and certainly go well beyond our unique ability to appose our thumbs to our fingers. What this brings up is the importance of the context in which the message is read. Read the code in a chimp embryo, and you build a chimp; read it in a human embryo, and you build a human.

This implies that the egg does not start life with a clean slate. Development of an organism may be more than just reading the message in DNA; there seems to be much more going on. We are able to look at the proteins that go into the making of the organism. At this level of comparison, the great differences between some organisms can be puzzling. They appear to have access to a similar pool of genes, as they have similar proteins; yet, they turn out very different.

To understand this, we need to remember that there are different organisational levels of DNA. Some genes code for proteins that we readily recognise, but others work behind the scene, and switch these genes on and off. When we look beyond those genes for structural proteins, we shall undoubtedly find that important differences exist.

We can envisage a case where the same DNA message has different developmental meanings. There is much more to the development of multicellular organisms than just reading the DNA code.

Just forty years ago, the world learned about DNA. The discovery of its central role in the cell followed soon after, but any hope that this knowledge would unveil the secret of life was dashed once the complicated behaviour and complex regulation of the gene became evident. Despite this, the amount of information we now have is astounding. We are beginning to put the pieces together into a wonderful whole, to at long last fulfil the dream of biologists to understand the very essence of life. Some general principles are emerging and with this comes the expectation that we might soon find out what makes us who we are.

We may then be able to read the mind of God.

FURTHER READING

Cohen J and Stewart I. *The Collapse of Chaos: Discovering Simplicity in a Complex World.* New York: Viking Penguin, 1994.

Crick F. *What Mad Pursuit: A Personal View of Scientific Discovery.* New York: Basic Books, 1988.

Daudel R. *The Realm of Molecules.* New York: McGraw-Hill, 1993.

Dulbecco R. *The Design of Life.* New Haven: Yale University Press, 1987.

Edey MA and Johanson DC. *Blueprints: Solving the Mystery of Evolution.* Boston: Little Brown and Company, 1989.

Gould SJ. *Wonderful Life. The Burgess Shale and the Nature of History.* New York: WW Norton and Company, 1989.

Jacob F. *The Logic of Life: A History of Heredity.* Princeton, New Jersey: Princeton University Press, 1993.

Jones S. *The Language of Genes.* New York: Anchor Books, 1993.

Kauffman S. *At Home in the Universe. The Search for the Laws of Self-organisation and Complexity.* New York: Oxford University Press, 1995.

Kordon C. *The Language of the Cell.* New York: McGraw-Hill, 1993.

Lewin R. *Complexity: Life at the Edge of Chaos.* New York: McMillan Publishing Company, 1992.

Olomucki M. *The Chemistry of Life.* New York: McGraw-Hill, 1993.

INDEX

A

AIDS 39,132
Amber 154
Antibiotic 123
Antibody 97,152
    diversity 98,105,106
    genes for 98,100-105
    gene mutations 106-107
    structure 98,**99**
Anticodon 50
Antigen 97
Atmospheric oxygen 47,65,67,68
Archaebacteria 66,67-71
    Halophiles 66
    Methanogens 66
    Thermophiles 66
    Thermoplasms 66
Atoms 22

B

Bacteria 60,63,64,65,67-71,155
    aerobic 67-68
    anaerobic 65,68
Bang, Oluf 111
Base pairing 30
Benzene 23,**24**
Binomial classification 59
Bishop, Michael 112,113

Blood cells 86-87
Bohr, Neils 22
Boole, George 144
Boolean networks 144-149
Boveri, Theodor 27, 111
Bowel cancer genes 119
BRCA genes 119
Breast cancer genes 119
Burgess Shale 74
Burkitt's lymphoma 115

C

Cambrian explosion 73-76,79,135
Cancer genes 39,109,112,113
    See also proto-oncogenes and
tumour suppressor genes
Cano, Raul 155
Carbon 21,23
Cech, Thomas 46
Cell **28**,122,123
    discovery 26
    DNA content 83
    number of types 83
    number of genes 140
    organisation 85
Cell cycle 79,**80**,89,118
Cell differentiation 86,88
Cell nucleus
27,30,34,61,72,90,156

Index

Index

P

Periodic table 21
Permian extinction 76
Photosynthesis 67,68
Phyla 75
Pikaia gracileus 75
Primogenitor 43,44,62,72-73
Prokaryotes 62,65,69,73
Protein 45,49,53,63,70,98,153,157
    enzymes 54,58,72
    gradients in embryo 88
    in metabolism 54-55
    regulatory 139
    synthesis 34,63,90
Proto-oncogene 113,114-116,118,119,133
    myc oncogene 115,116
    ras oncogene 116,130
    src oncogene 112,113,114
Protovirus hypothesis 39
Provirus 56,112,125,127,132,133
Provirus hypothesis 38

R

Ramayana 110
Recessive traits 16
Repair gene 118
Retinoblastoma 117
Retrovirus 38,39,41,56,112,113,124,127,131
Reverse transcriptase 39,112,124
Ribosome 36, 49-50, 54, 57, 64, 89, 153
Ribozyme 46,47,132-133
RNA 29
    catalytic RNA 46,48,57,72, *see also* ribozyme
    first living molecule 46,47-48,55,72
    messenger RNA 34,36,103

origin 47
ribosomal RNA 36,50,64,65,89
    transfer RNA 50-51,**52**
RNA world 49,56
RNP world 49,56
Rosenberg, Steven 130
Rous, Francis 37,38,111
Roux, Wilhelm 78

S

Sarcoma 37,111
Schleiden, Mathias 110,122
Schwann, Theodor 110, 122
Second adaptor 51,53
Sickle cell anaemia 128
Stem cell 86-87
Stromatolite 62
Szostak, Jack 47

T

Temin, Howard 38,39
Tetrahymena 46,48
Thalassaemia 128
Tonegawa, Susumu 101
Tree of Life 64,65,70,**71**,72
Tumour necrosis factor 129
Tumour suppressor gene 118,119
    p53 gene 118,130
    rb gene 117

U

Urey, Harold 47

V

Varmus, Harold 112,113
Virchow, Rudolph 110,122

www.ingramcontent.com/pod-product-compliance
Lightning Source LLC
Chambersburg PA
CBHW060031210326
41520CB00009B/1083